All inquiries should be addressed to:

www.EffortlessMath.com

www.EffortlessMath.com

ISBN-13: 978-1-970036-08-4

ISBN-10: 1-970036-08-7

Published by: Effortless Math Education

www.EffortlessMath.com

Description

HSPT Mathematics Prep 2019 provides students with the confidence and math skills they need to succeed on the HSPT Math, building a solid foundation of basic Math topics with abundant exercises for each topic. It is designed to address the needs of HSPT test takers who must have a working knowledge of basic Math.

This comprehensive book with over 2,500 sample questions and 2 complete HSPT tests is all you need to fully prepare for the HSPT Math. It will help you learn everything you need to ace the math section of the HSPT.

Effortless Math unique study program provides you with an in-depth focus on the math portion of the exam, helping you master the math skills that students find the most troublesome.

This book contains most common sample questions that are most likely to appear in the mathematics section of the HSPT.

Inside the pages of this comprehensive HSPT Math book, students can learn basic math operations in a structured manner with a complete study program to help them understand essential math skills. It also has many exciting features, including:

- Dynamic design and easy-to-follow activities
- A fun, interactive and concrete learning process
- Targeted, skill-building practices
- Fun exercises that build confidence
- Math topics are grouped by category, so you can focus on the topics you struggle on
- All solutions for the exercises are included, so you will always find the answers
- 2 Complete HSPT Math Practice Tests that reflect the format and question types on HSPT

HSPT Mathematics 2019 is an incredibly useful tool for those who want to review all topics being covered on the HSPT test. It efficiently and effectively reinforces learning outcomes through engaging questions and repeated practice, helping you to quickly master basic Math skills.

About the Author

Reza Nazari is the author of more than 100 Math learning books including:
– **Math and Critical Thinking Challenges:** For the Middle and High School Student
– **GED Math in 30 Days**
– **ASVAB Math Workbook 2018 - 2019**
– **Effortless Math Education Workbooks**
– **and many more Mathematics books …**

Reza is also an experienced Math instructor and a test–prep expert who has been tutoring students since 2008. Reza is the founder of Effortless Math Education, a tutoring company that has helped many students raise their standardized test scores—and attend the colleges of their dreams. Reza provides an individualized custom learning plan and the personalized attention that makes a difference in how students view math.

You can contact Reza via email at:
reza@EffortlessMath.com

Find Reza's professional profile at:
goo.gl/zoC9rJ

Contents

Chapter 1: Whole Numbers

Topics that you'll learn in this chapter:

- ✓ Place Value
- ✓ Rounding
- ✓ Whole Number Addition and Subtraction
- ✓ Whole Number Multiplication and Division
- ✓ Rounding and Estimates
- ✓ Comparing Numbers

"If people do not believe that mathematics is simple, it is only because they do not realize how complicated life is." — John von Neumann

Place Value

Helpful	The value of the place, or position, of a digit in a number.	**Example:**
	For the number 3,684.26	In 456, the 5 is in
Hints		"tens" position.

Decimal Place Value Chart

Millions	Hundred thousands	Ten thousands	Thousands	Hundreds	Tens	Ones	Decimal point	Tenths	Hundredths	Thousandths	Ten-thousandths	Hundred thousandths	Millionths
			3	6	8	4	.	2	6				

✍ Write each number in expanded form.

1) Thirty–five   30 + 5

2) Sixty–seven   ___ + ___

3) Forty–two   ___ + ___

4) Eighty–nine   ___ + ___

5) Ninety–one   ___ + ___

✍ Circle the correct one.

6)   The 2 in 72 is in the   ones place   tens place   hundreds place

7)   The 6 in 65 is in the   ones place   tens place   hundreds place

8)   The 2 in 342 is in the   ones place   tens place   hundreds place

9)   The 5 in 450 is in the   ones place   tens place   hundreds place

10)   The 3 in 321 is in the   ones place   tens place   hundreds place

Rounding

Helpful *Hints*	–	Rounding is putting a number up or down to the nearest whole number or the nearest hundred, etc.	**Example:** 64 rounded to the nearest ten is 60, because 64 is closer to 60 than to 70.

✎ *Round each number to the underlined place value.*

1) 9̲72

2) 2,9̲95

3) 36̲4

4) 8̲1

5) 5̲5

6) 33̲4

7) 1,2̲03

8) 9.5̲7

9) 7.4̲84

10) 9.1̲4

11) 3̲9

12) 9̲,123

13) 3,45̲2

14) 5̲69

15) 1,2̲30

16) 9̲8

17) 9̲3

18) 3̲7

19) 49̲3

20) 2,9̲23

21) 9̲,845

22) 55̲5

23) 9̲39

24) 6̲9

Whole Number Addition and Subtraction

Helpful	1– Line up the numbers.	**Example:**
	2– Start with the unit place. (ones place)	$231 + 120 = 351$
Hints	3– Regroup if necessary.	$292 - 90 = 202$
	4– Add or subtract the tens place.	
	5– Continue with other digits.	

✎ *Solve.*

1) A school had 891 students last year. If all last year students and 338 new students have registered for this year, how many students will there be in total?

2) Alice has just started her first job after graduating from college. Her yearly income is $33,000 per year. Alice's father income is $56,000 per year and her mother's income is $49,000. What is yearly income of Alice and her parent altogether?

3) Tom had $895 dollars in his saving account. He gave $235 dollars to his sister, Lisa. How much money does he have left?

4) Emily has 830 marbles, Daniel has 970 marbles, and Ethan has 230 marbles less than Daniel. How many marbles do they have in all?

✎ *Find the missing number.*

5) $890 - \ldots\ldots\ldots\ldots = 300$

6) $1000 - \ldots\ldots\ldots\ldots = 200$

7) $\ldots\ldots\ldots\ldots - 4000 = 92000$

8) $60000 - 51000 = \ldots\ldots\ldots\ldots$

9) $3400 - \ldots\ldots\ldots\ldots = 3200$

10) $33000 - 5000 = \ldots\ldots\ldots\ldots$

Whole Number Multiplication and Division

Helpful *Hints*	**Multiplication:** – Learn the times tables first! – For multiplication, line up the numbers you are multiplying. – Start with the ones place. – Continue with other digits – A typical division problem: Dividend ÷ Divisor = Quotient **Division:** – In division, we want to find how many times a number (divisor) is contained in another number (dividend). – The result in a division problem is the quotient.	**Example:** $200 \times 90 = 18{,}000$ $18{,}000 \div 90 = 200$

✍ Multiply and divided.

1) $340 \div 8 =$

2) $1800 \div 20 =$

3) $50000 \div 10 =$

4) $966 \div 30 =$

5) $201 \times 20 =$

6) $400 \times 50 =$

7) $400 \times 90 =$

8) $888 \times 90 =$

9) $80 \times 80 =$

10) $122 \times 12 =$

11) $609 \times 8 =$

12) $220 \times 12 =$

13) A group of 235 students has collected $8,565 for charity during last month. They decided to split the money evenly among 5 charities. How much will each charity receive?

14) Maria and her two brothers have 9 boxes of crayons. Each box contains 56 crayons. How many crayons do Maria and her two brothers have?

Rounding and Estimates

Helpful *Hints*	– Rounding and estimating are math strategies used for approximating a number.	**Example:**
	– To estimate means to make a rough guess or calculation.	$73 + 69 \approx 140$
	– To round means to simplify a known number by scaling it slightly up or down.	

✍ ***Estimate the sum by rounding each added to the nearest ten.***

1) 55 + 9

2) 25 + 12

3) 83 + 7

4) 32 + 37

5) 13 + 74

6) 34 + 11

7) 39 + 77

8) 25 + 4

9) 61 + 73

10) 64 + 59

11) 14 + 68

12) 82 + 12

13) 43 + 66

14) 45 + 65

15) 553 + 232

16) 418 + 846

17) 582 + 277

18) 2771 + 1651

19) 7436 + 3575

20) 1542 + 8738

21) 3843 + 6579

22) 4722 + 8186

23) 2419 + 7224

24) 6768 + 3169

Comparing Numbers

Helpful	Comparing:	Example:
	Equal to =	
Hints	Less than <	56 > 35
	Greater than >	
	Greater than or equal ≥	
	Less than or equal ≤	

✍ **Use > = <.**

1) 35   67

2) 89   56

3) 56   35

4) 27   56

5) 34   34

6) 28   45

7) 89   67

8) 90   56

9) 94   98

10) 48   23

11) 24   54

12) 89   89

13) 50   30

14) 20   20

✍ **Use less than, equal to or greater than.**

15) 23 _____ 34

16) 89 _____ 98

17) 45 _____ 25

18) 34 _____ 32

19) 91 _____ 91

20) 57 _____ 55

21) 85 _____ 78

22) 56 _____ 43

23) 34 _____ 34

24) 92 _____ 98

25) 38 _____ 46

26) 67 _____ 58

27) 88 _____ 69

28) 23 _____ 34

Answers of Worksheets – Chapter 1

Place Value

1) 30 + 5
2) 60 + 7
3) 40 + 2
4) 80 + 9

5) 90 + 1
6) ones place
7) tens place
8) ones place

9) tens place
10) hundreds place

Rounding

1) 1000
2) 3000
3) 360
4) 80
5) 60
6) 330
7) 1200
8) 9.6

9) 7.5
10) 9.1
11) 40
12) 9000
13) 3,450
14) 600
15) 1,200
16) 100

17) 90
18) 40
19) 490
20) 2,900
21) 10,000
22) 560
23) 900
24) 70

Whole Number Addition and Subtraction

1) 1229
2) 138000
3) 660
4) 2540

5) 590
6) 800
7) 96000
8) 9000

9) 200
10) 28000

Whole Number Multiplication and Division

1) 42.5
2) 90
3) 5000
4) 32.2
5) 4020

6) 20000
7) 36000
8) 79920
9) 6400
10) 1464

11) 4872
12) 2640
13) 1713
14) 504

Rounding and Estimates

1) 70

2) 40

3) 90

4) 70

5) 37

6) 40

7) 120

8) 30

9) 130

10) 120

11) 80

12) 90

13) 110

14) 120

15) 780

16) 1270

17) 860

18) 4420

19) 11020

20) 10280

21) 10420

22) 12910

23) 9640

24) 9940

Comparing Numbers

1) 35 < 67

2) 89 > 56

3) 56 > 35

4) 27< 56

5) 34 = 34

6) 28 < 45

7) 89 > 67

8) 90 > 56

9) 94 < 98

10) 48 > 23

11) 24 < 54

12) 89 = 89

13) 50 > 30

14) 20 = 20

15) 23 less than 34

16) 89 less than 98

17) 45 greater than 25

18) 34 greater than 32

19) 91 equal to 91

20) 57 greater than 55

21) 85 greater than 78

22) 56 greater than 43

23) 34 equal to 34

24) 92 less than 98

25) 38 less than 46

26) 67 greater than 58

27) 88 greater than 69

28) 23 less than 34

Chapter 2: Fractions and Decimals

Topics that you'll learn in this chapter:

- ✓ Simplifying Fractions
- ✓ Adding and Subtracting Fractions
- ✓ Multiplying and Dividing Fractions
- ✓ Adding Mixed Numbers
- ✓ Subtract Mixed Numbers
- ✓ Multiplying Mixed Numbers
- ✓ Dividing Mixed Numbers
- ✓ Comparing Decimals
- ✓ Rounding Decimals

- ✓ Adding and Subtracting Decimals
- ✓ Multiplying and Dividing Decimals
- ✓ Converting Between Fractions, Decimals and Mixed Numbers
- ✓ Factoring Numbers
- ✓ Greatest Common Factor
- ✓ Least Common Multiple
- ✓ Divisibility Rules

"A Man is like a fraction whose numerator is what he is and whose denominator is what he thinks of himself. The larger the denominator, the smaller the fraction." –Tolstoy

Simplifying Fractions

Helpful	– Evenly divide both the top and bottom of the fraction by 2, 3, 5, 7, … etc.	**Example:**
Hints	– Continue until you can't go any further.	$\frac{4}{12} = \frac{2}{6} = \frac{1}{3}$

✎ *Simplify the fractions.*

1) $\frac{22}{36}$

2) $\frac{8}{10}$

3) $\frac{12}{18}$

4) $\frac{6}{8}$

5) $\frac{13}{39}$

6) $\frac{5}{20}$

7) $\frac{16}{36}$

8) $\frac{18}{36}$

9) $\frac{20}{50}$

10) $\frac{6}{54}$

11) $\frac{45}{81}$

12) $\frac{21}{28}$

13) $\frac{35}{56}$

14) $\frac{52}{64}$

15) $\frac{13}{65}$

16) $\frac{44}{77}$

17) $\frac{21}{42}$

18) $\frac{15}{36}$

19) $\frac{9}{24}$

20) $\frac{20}{80}$

21) $\frac{25}{45}$

Adding and Subtracting Fractions

Helpful	− For "like" fractions (fractions with the same denominator), add or subtract the numerators and write the answer over the common denominator.
Hints	− Find equivalent fractions with the same denominator before you can add or subtract fractions with different denominators.

− Adding and Subtracting with the same denominator:

$$\frac{a}{b} + \frac{c}{b} = \frac{a+c}{b}$$
$$\frac{a}{b} - \frac{c}{b} = \frac{a-c}{b}$$

− Adding and Subtracting fractions with different denominators:

$$\frac{a}{b} + \frac{c}{d} = \frac{ad+cb}{bd}$$
$$\frac{a}{b} - \frac{c}{d} = \frac{ad-cb}{bd}$$

✎Add fractions.

1) $\frac{2}{3} + \frac{1}{2}$

2) $\frac{3}{5} + \frac{1}{3}$

3) $\frac{5}{6} + \frac{1}{2}$

4) $\frac{7}{4} + \frac{5}{9}$

5) $\frac{2}{5} + \frac{1}{5}$

6) $\frac{3}{7} + \frac{1}{2}$

7) $\frac{3}{4} + \frac{2}{5}$

8) $\frac{2}{3} + \frac{1}{5}$

9) $\frac{16}{25} + \frac{3}{5}$

✎Subtract fractions.

10) $\frac{4}{5} - \frac{2}{5}$

11) $\frac{3}{5} - \frac{2}{7}$

12) $\frac{1}{2} - \frac{1}{3}$

13) $\frac{8}{9} - \frac{3}{5}$

14) $\frac{3}{7} - \frac{3}{14}$

15) $\frac{4}{15} - \frac{1}{10}$

16) $\frac{3}{4} - \frac{13}{18}$

17) $\frac{5}{8} - \frac{2}{5}$

18) $\frac{1}{2} - \frac{1}{9}$

Multiplying and Dividing Fractions

Helpful	– **Multiplying fractions:** multiply the top numbers and multiply the bottom numbers.	**Example:**
	– **Dividing fractions:** Keep, Change, Flip	$\frac{a}{b} \times \frac{c}{d} = \frac{a \times c}{b \times d}$
Hints	Keep first fraction, change division sign to multiplication, and flip the numerator and denominator of the second fraction. Then, solve!	$\frac{a}{b} \div \frac{c}{d} = \frac{a}{b} \times \frac{d}{c} = \frac{ad}{bc}$

✍ *Multiplying fractions. Then simplify.*

1) $\frac{1}{5} \times \frac{2}{3}$

2) $\frac{3}{4} \times \frac{2}{3}$

3) $\frac{2}{5} \times \frac{3}{7}$

4) $\frac{3}{8} \times \frac{1}{3}$

5) $\frac{3}{5} \times \frac{2}{5}$

6) $\frac{7}{9} \times \frac{1}{3}$

7) $\frac{2}{3} \times \frac{3}{8}$

8) $\frac{1}{4} \times \frac{1}{3}$

9) $\frac{5}{7} \times \frac{7}{12}$

✍ *Dividing fractions.*

10) $\frac{2}{9} \div \frac{1}{4}$

11) $\frac{1}{2} \div \frac{1}{3}$

12) $\frac{6}{11} \div \frac{3}{4}$

13) $\frac{11}{14} \div \frac{1}{10}$

14) $\frac{3}{5} \div \frac{5}{9}$

15) $\frac{1}{2} \div \frac{1}{2}$

16) $\frac{3}{5} \div \frac{1}{5}$

17) $\frac{12}{21} \div \frac{3}{7}$

18) $\frac{5}{14} \div \frac{9}{10}$

Adding Mixed Numbers

Helpful	Use the following steps for both adding and subtracting mixed numbers.	**Example:**
Hints	– Find the Least Common Denominator (LCD) – Find the equivalent fractions for each mixed number. – Add fractions after finding common denominator. – Write your answer in lowest terms.	$1\frac{3}{4} + 2\frac{3}{8} = 4\frac{1}{8}$

✎ **Add.**

1) $4\frac{1}{2} + 5\frac{1}{2}$

2) $2\frac{3}{8} + 3\frac{1}{8}$

3) $6\frac{1}{5} + 3\frac{2}{5}$

4) $1\frac{1}{3} + 2\frac{2}{3}$

5) $5\frac{1}{6} + 5\frac{1}{2}$

6) $3\frac{1}{3} + 1\frac{1}{3}$

7) $1\frac{10}{11} + 1\frac{1}{3}$

8) $2\frac{3}{6} + 1\frac{1}{2}$

9) $5\frac{3}{5} + 5\frac{1}{5}$

10) $7 + \frac{1}{5}$

11) $1\frac{5}{7} + \frac{1}{3}$

12) $2\frac{1}{4} + 1\frac{1}{2}$

Subtract Mixed Numbers

Helpful	Use the following steps for both adding and subtracting mixed numbers.	**Example:**
Hints	Find the Least Common Denominator (LCD) – Find the equivalent fractions for each mixed number. – Add or subtract fractions after finding common denominator. – Write your answer in lowest terms.	$5\dfrac{2}{3} - 3\dfrac{2}{7} = 2\dfrac{8}{21}$

✎ **Subtract.**

1) $4\dfrac{1}{2} - 3\dfrac{1}{2}$

2) $3\dfrac{3}{8} - 3\dfrac{1}{8}$

3) $6\dfrac{3}{5} - 5\dfrac{1}{5}$

4) $2\dfrac{1}{3} - 1\dfrac{2}{3}$

5) $6\dfrac{1}{6} - 5\dfrac{1}{2}$

6) $3\dfrac{1}{3} - 1\dfrac{1}{3}$

7) $2\dfrac{10}{11} - 1\dfrac{1}{3}$

8) $2\dfrac{1}{2} - 1\dfrac{1}{2}$

9) $6\dfrac{3}{5} - 2\dfrac{1}{5}$

10) $7\dfrac{2}{5} - 1\dfrac{1}{5}$

11) $2\dfrac{5}{7} - 1\dfrac{1}{3}$

12) $2\dfrac{1}{4} - 1\dfrac{1}{2}$

Multiplying Mixed Numbers

Helpful	1- Convert the mixed numbers to improper fractions.	**Example:**
	2- Multiply fractions and simplify if necessary.	$2\dfrac{1}{3} \times 5\dfrac{3}{7} =$
Hints	$a\dfrac{c}{b} = a + \dfrac{c}{b} = \dfrac{ab+c}{b}$	$\dfrac{7}{3} \times \dfrac{38}{7} = \dfrac{38}{3} = 12\dfrac{2}{3}$

✎ **Find each product.**

1) $1\dfrac{2}{3} \times 1\dfrac{1}{4}$

2) $1\dfrac{3}{5} \times 1\dfrac{2}{3}$

3) $1\dfrac{2}{3} \times 3\dfrac{2}{7}$

4) $4\dfrac{1}{8} \times 1\dfrac{2}{5}$

5) $2\dfrac{2}{5} \times 3\dfrac{1}{5}$

6) $1\dfrac{1}{3} \times 1\dfrac{2}{3}$

7) $1\dfrac{5}{8} \times 2\dfrac{1}{2}$

8) $3\dfrac{2}{5} \times 2\dfrac{1}{5}$

9) $2\dfrac{2}{3} \times 4\dfrac{1}{4}$

10) $2\dfrac{3}{5} \times 1\dfrac{2}{4}$

11) $1\dfrac{1}{3} \times 1\dfrac{1}{4}$

12) $3\dfrac{2}{5} \times 1\dfrac{1}{5}$

Dividing Mixed Numbers

| *Helpful* *Hints* | 1- Convert the mixed numbers to improper fractions.
2- Divide fractions and simplify if necessary.

$$a\frac{c}{b} = a + \frac{c}{b} = \frac{ab+c}{b}$$ | **Example:**
$$10\frac{1}{2} \div 5\frac{3}{5} =$$
$$\frac{21}{2} \div \frac{28}{5} = \frac{21}{2} \times \frac{5}{28} = \frac{105}{56}$$
$$= 1\frac{7}{8}$$ |

✎ **Find each quotient.**

1) $2\frac{1}{5} \div 2\frac{1}{2}$

2) $2\frac{3}{5} \div 1\frac{1}{3}$

3) $3\frac{1}{6} \div 4\frac{2}{3}$

4) $1\frac{2}{3} \div 3\frac{1}{3}$

5) $4\frac{1}{8} \div 2\frac{2}{4}$

6) $3\frac{1}{2} \div 2\frac{3}{5}$

7) $3\frac{5}{9} \div 1\frac{2}{5}$

8) $2\frac{2}{7} \div 1\frac{1}{2}$

9) $3\frac{1}{5} \div 1\frac{1}{2}$

10) $4\frac{3}{5} \div 2\frac{1}{3}$

11) $6\frac{1}{6} \div 1\frac{2}{3}$

12) $2\frac{2}{3} \div 1\frac{1}{3}$

Comparing Decimals

Helpful	-	**Decimals:** is a fraction written in a special form. For example, instead of writing $\frac{1}{2}$ you can write 0.5.	**Example:**
Hints	-	**For comparing:** Equal to = Less than < Greater than > Greater than or equal ≥ Less than or equal ≤	2.67 > 0.267

✎*Write the correct comparison symbol (>, < or =).*

1) 1.25   2.3

2) 0.5   0.23

3) 3.2   3.2

4) 4.58   45.8

5) 2.75   0.275

6) 5.2   5

7) 3.1   0.31

8) 6.33   0.733

9) 8   0.8

10) 4.56   0.456

11) 1.12   1.14

12) 2.77   2.78

13) 6.08   6.11

14) 1.11   0.211

15) 2.6   2.55

16) 1.24   1.25

17) 5.52   0.552

18) 0.33   0.033

19) 14.4   14.4

20) 0.05   0.50

21) 0.59   0.7

22) 0.5   0.05

23) 0.90   0.9

24) 0.27   0.4

Rounding Decimals

Helpful *Hints*	We can round decimals to a certain accuracy or number of decimal places. This is used to make calculation easier to do and results easier to understand, when exact values are not too important. First, you'll need to remember your place values:	**Example:**

12.4567

1: tens	2: ones	4: tenths
5: hundredths	6: thousandths	7: tens thousandths

$\underline{6}.37 = 6$

✎ *Round each decimal number to the nearest place indicated.*

1) 0.2$\underline{3}$	9) 1.6$\underline{2}$9	17) 70.$\underline{7}$8
2) 4.$\underline{0}$4	10) 6.$\underline{3}$959	18) 61$\underline{5}$.755
3) 5.$\underline{6}$23	11) $\underline{1}$.9	19) 1$\underline{6}$.4
4) 0.$\underline{2}$66	12) $\underline{5}$.2167	20) 9$\underline{5}$.81
5) $\underline{6}$.37	13) 5.$\underline{8}$63	21) $\underline{2}$.408
6) 0.8$\underline{8}$	14) 8.$\underline{5}$4	22) 7$\underline{6}$.3
7) 8.2$\underline{4}$	15) 80.$\underline{6}$9	23) 116.$\underline{5}$14
8) $\underline{7}$.0760	16) 6$\underline{5}$.85	24) 8.$\underline{0}$6

Adding and Subtracting Decimals

Helpful Hints	1– Line up the numbers.	Example:
	2– Add zeros to have same number of digits for both numbers.	$\begin{array}{r} 16.18 \\ -\ 13.45 \\ \hline 2.73 \end{array}$
	3– Add or Subtract using column addition or subtraction.	

✍ *Add and subtract decimals.*

1) $\begin{array}{r} 15.14 \\ -\ 12.18 \\ \hline \end{array}$

3) $\begin{array}{r} 82.56 \\ +\ 12.28 \\ \hline \end{array}$

5) $\begin{array}{r} 90.37 \\ +\ 56.97 \\ \hline \end{array}$

2) $\begin{array}{r} 65.72 \\ +\ 43.67 \\ \hline \end{array}$

4) $\begin{array}{r} 34.18 \\ -\ 23.45 \\ \hline \end{array}$

6) $\begin{array}{r} 45.78 \\ -\ 23.39 \\ \hline \end{array}$

✍ *Solve.*

7) ____ + 1.3 = 4.8

10) 6.9 + ____ = 16.4

8) 4.2 + ____ = 11.6

11) ____ + 5.1 = 8.6

9) 9.9 + ____ = 16

12) ____ + 7.9 = 15.2

Multiplying and Dividing Decimals

Helpful	**For Multiplication:**
	– Set up and multiply the numbers as you do with whole numbers.
Hints	– Count the total number of decimal places in both of the factors.
	– Place the decimal point in the product.
	For Division:
	– If the divisor is not a whole number, move decimal point to right to make it a whole number. Do the same for dividend.
	– Divide similar to whole numbers.

✎ Find each product.

1) $\begin{array}{r} 4.5 \\ \times\ 1.6 \\ \hline \end{array}$

2) $\begin{array}{r} 7.7 \\ \times\ 9.9 \\ \hline \end{array}$

3) $\begin{array}{r} 2.6 \\ \times\ 1.5 \\ \hline \end{array}$

4) $\begin{array}{r} 8.9 \\ \times\ 9.7 \\ \hline \end{array}$

5) $\begin{array}{r} 15.1 \\ \times\ 12.6 \\ \hline \end{array}$

6) $\begin{array}{r} 6.9 \\ \times\ 3.3 \\ \hline \end{array}$

7) $\begin{array}{r} 5.7 \\ \times\ 7.8 \\ \hline \end{array}$

8) $\begin{array}{r} 98.20 \\ \times\ 100 \\ \hline \end{array}$

9) $\begin{array}{r} 23.99 \\ \times\ 1000 \\ \hline \end{array}$

✎ Find each quotient.

10) $9.2 \div 3.6$

11) $27.6 \div 3.8$

12) $12.6 \div 4.7$

13) $6.5 \div 8.1$

14) $1.4 \div 10$

15) $3.6 \div 100$

16) $4.24 \div 10$

17) $14.6 \div 100$

18) $1.8 \div 1000$

Converting Between Fractions, Decimals and Mixed Numbers

Helpful Hints	**Fraction to Decimal:**
	– Divide the top number by the bottom number.
	Decimal to Fraction:
	– Write decimal over 1.
	– Multiply both top and bottom by 10 for every digit on the right side of the decimal point.
	– Simplify.

✍ *Convert fractions to decimals.*

1) $\dfrac{9}{10}$ 　　　　　4) $\dfrac{2}{5}$ 　　　　　7) $\dfrac{12}{10}$

2) $\dfrac{56}{100}$ 　　　　5) $\dfrac{3}{9}$ 　　　　　8) $\dfrac{8}{5}$

3) $\dfrac{3}{4}$ 　　　　　6) $\dfrac{40}{50}$ 　　　　9) $\dfrac{69}{10}$

✍ *Convert decimal into fraction or mixed numbers.*

10) 0.3 　　　　　14) 0.8 　　　　　18) 0.08

11) 4.5 　　　　　15) 0.25 　　　　19) 0.45

12) 2.5 　　　　　16) 0.14 　　　　20) 2.6

13) 2.3 　　　　　17) 0.2 　　　　　21) 5.2

Factoring Numbers

Helpful	-	Factoring numbers means to break the numbers into their prime factors.	**Example:**
Hints	-	First few prime numbers: 2, 3, 5, 7, 11, 13, 17, 19	$12 = 2 \times 2 \times 3$

✍️ **List all positive factors of each number.**

1) 68

2) 56

3) 24

4) 40

5) 86

6) 78

7) 50

8) 98

9) 45

10) 26

11) 54

12) 28

13) 55

14) 85

15) 48

✍️ **List the prime factorization for each number.**

16) 50

17) 25

18) 69

19) 21

20) 45

21) 68

22) 26

23) 86

24) 93

Greatest Common Factor

		Example:
Helpful	- List the prime factors of each number.	
	- Multiply common prime factors.	$200 = 2 \times 2 \times 2 \times 5 \times 5$
Hints		$60 = 2 \times 2 \times 3 \times 5$
		GCF (200, 60) = $2 \times 2 \times 5 = 20$

✍️ **Find the GCF for each number pair.**

1) 20, 30

2) 4, 14

3) 5, 45

4) 68, 12

5) 5, 12

6) 15, 27

7) 3, 24

8) 34, 6

9) 4, 10

10) 5, 3

11) 6, 16

12) 30, 3

13) 24, 28

14) 70, 10

15) 45, 8

16) 90, 35

17) 78, 34

18) 55, 75

19) 60, 72

20) 100, 78

21) 30, 40

Least Common Multiple

| Helpful Hints | - Find the GCF for the two numbers.
- Divide that GCF into either number.
- Take that answer and multiply it by the other number. | **Example:**

LCM (200, 60):

GCF is 20

$200 \div 20 = 10$

$10 \times 60 = 600$ |

✍ *Find the LCM for each number pair.*

1) 4, 14

2) 5, 15

3) 16, 10

4) 4, 34

5) 8, 3

6) 12, 24

7) 9, 18

8) 5, 6

9) 8, 19

10) 9, 21

11) 19, 29

12) 7, 6

13) 25, 6

14) 4, 8

15) 30, 10, 50

16) 18, 36, 27

17) 12, 8, 18

18) 8, 18, 4

19) 26, 20, 30

20) 10, 4, 24

21) 15, 30, 45

Divisibility Rules

| Helpful Hints | - | Divisibility means that a number can be divided by other numbers evenly. | **Example:** 24 is divisible by 6, because 24 ÷ 6 = 4 |

✍ *Use the divisibility rules to find the factors of each number.*

8

<u>2</u> 3 <u>4</u> 5 6 7 <u>8</u> 9 10

1) 16

2 3 4 5 6 7 8 9 10

2) 10

2 3 4 5 6 7 8 9 10

3) 15

2 3 4 5 6 7 8 9 10

4) 28

2 3 4 5 6 7 8 9 10

5) 36

2 3 4 5 6 7 8 9 10

6) 15

2 3 4 5 6 7 8 9 10

7) 27

2 3 4 5 6 7 8 9 10

8) 70

2 3 4 5 6 7 8 9 10

9) 57

2 3 4 5 6 7 8 9 10

10) 102

2 3 4 5 6 7 8 9 10

11) 144

2 3 4 5 6 7 8 9 10

12) 75

2 3 4 5 6 7 8 9 10

Answers of Worksheets – Chapter 2

Simplifying Fractions

1) $\dfrac{11}{18}$

2) $\dfrac{4}{5}$

3) $\dfrac{2}{3}$

4) $\dfrac{3}{4}$

5) $\dfrac{1}{3}$

6) $\dfrac{1}{4}$

7) $\dfrac{4}{9}$

8) $\dfrac{1}{2}$

9) $\dfrac{2}{5}$

10) $\dfrac{1}{9}$

11) $\dfrac{5}{9}$

12) $\dfrac{3}{4}$

13) $\dfrac{5}{8}$

14) $\dfrac{13}{16}$

15) $\dfrac{1}{5}$

16) $\dfrac{4}{7}$

17) $\dfrac{1}{2}$

18) $\dfrac{5}{12}$

19) $\dfrac{3}{8}$

20) $\dfrac{1}{4}$

21) $\dfrac{5}{9}$

Adding and Subtracting Fractions

1) $\dfrac{7}{6}$

2) $\dfrac{14}{15}$

3) $\dfrac{4}{3}$

4) $\dfrac{83}{36}$

5) $\dfrac{3}{5}$

6) $\dfrac{13}{14}$

7) $\dfrac{23}{20}$

8) $\dfrac{13}{15}$

9) $\dfrac{31}{25}$

10) $\dfrac{2}{5}$

11) $\dfrac{11}{35}$

12) $\dfrac{1}{6}$

13) $\dfrac{13}{45}$

14) $\dfrac{3}{14}$

15) $\dfrac{1}{6}$

16) $\dfrac{1}{36}$

17) $\dfrac{9}{40}$

18) $\dfrac{7}{18}$

Multiplying and Dividing Fractions

1) $\dfrac{2}{15}$

2) $\dfrac{1}{2}$

3) $\dfrac{6}{35}$

4) $\dfrac{1}{8}$

5) $\dfrac{6}{25}$

6) $\dfrac{7}{27}$

7) $\dfrac{1}{4}$

8) $\dfrac{1}{12}$

9) $\dfrac{5}{12}$

10) $\dfrac{8}{9}$

11) $\dfrac{3}{2}$

12) $\dfrac{8}{11}$

13) $\dfrac{55}{7}$

14) $\dfrac{27}{25}$

15) 1

16) 3

17) $\dfrac{4}{3}$

18) $\dfrac{25}{63}$

Adding Mixed Numbers

1) 10

2) $5\dfrac{1}{2}$

3) $9\dfrac{3}{5}$

4) 4

5) $10\dfrac{2}{3}$

6) $4\dfrac{2}{3}$

7) $3\dfrac{8}{33}$

8) 4

9) $10\dfrac{4}{5}$

10) $7\dfrac{1}{5}$

11) $2\dfrac{1}{21}$

12) $3\dfrac{3}{4}$

Subtract Mixed Numbers

1) 1

2) $\dfrac{1}{4}$

3) $1\dfrac{2}{5}$

4) $\dfrac{2}{3}$

5) $\dfrac{2}{3}$

6) 2

7) $1\dfrac{19}{33}$

8) 1

9) $4\dfrac{2}{5}$

10) $6\dfrac{1}{5}$

11) $1\dfrac{8}{21}$

12) $\dfrac{3}{4}$

Multiplying Mixed Numbers

1) $2\frac{1}{12}$

2) $2\frac{2}{3}$

3) $5\frac{10}{21}$

4) $5\frac{31}{40}$

5) $7\frac{17}{25}$

6) $2\frac{2}{9}$

7) $4\frac{1}{16}$

8) $7\frac{12}{25}$

9) $11\frac{1}{3}$

10) $3\frac{9}{10}$

11) $1\frac{2}{3}$

12) $4\frac{2}{25}$

Dividing Mixed Numbers

1) $\frac{22}{25}$

2) $1\frac{19}{20}$

3) $\frac{19}{28}$

4) $\frac{1}{2}$

5) $1\frac{13}{20}$

6) $1\frac{9}{26}$

7) $2\frac{34}{63}$

8) $1\frac{11}{21}$

9) $2\frac{2}{15}$

10) $1\frac{34}{35}$

11) $3\frac{7}{10}$

12) 2

Comparing Decimals

1) $1.25 < 2.3$

2) $0.5 > 0.23$

3) $3.2 = 3.2$

4) $4.58 < 45.8$

5) $2.75 > 0.275$

6) $5.2 > 5$

7) $3.1 > 0.31$

8) $6.33 > 0.733$

9) $8 > 0.8$

10) $4.56 > 0.456$

11) $1.12 < 1.14$

12) $2.77 < 2.78$

13) $6.08 < 6.11$

14) $1.11 > 0.211$

15) $2.6 > 2.55$

16) $1.24 < 1.25$

17) $5.52 > 0.552$

18) $0.33 > 0.033$

19) $14.4 = 14.4$

20) $0.05 < 0.50$

21) $0.59 < 0.7$

22) $0.5 > 0.05$

23) $0.90 = 0.9$

24) $0.27 < 0.4$

Rounding Decimals

1) 0.2
2) 4.0
3) 5.6
4) 0.3
5) 6
6) 0.9
7) 8.2
8) 7
9) 1.63
10) 6.4
11) 2
12) 5
13) 5.9
14) 8.5
15) 81
16) 66
17) 70.8
18) 616
19) 16
20) 96
21) 2
22) 76
23) 116.5
24) 8.1

Adding and Subtracting Decimals

1) 2.96
2) 109.39
3) 94.84
4) 10.73
5) 147.34
6) 22.39
7) 3.5
8) 7.4
9) 6.1
10) 9.5
11) 3.5
12) 7.3

Multiplying and Dividing Decimals

1) 7.2
2) 76.23
3) 3.9
4) 86.33
5) 190.26
6) 22.77
7) 44.46
8) 9820
9) 23990
10) 2.5555…
11) 7.2631…
12) 2.6808…
13) 0.8024…
14) 0.14
15) 0.036
16) 0.424
17) 0.146
18) 0.0018

Converting Between Fractions, Decimals and Mixed Numbers

1) 0.9
2) 0.56
3) 0.75
4) 0.4
5) 0.333…
6) 0.8
7) 1.2
8) 1.6
9) 6.9
10) $\frac{3}{10}$
11) $4\frac{1}{2}$
12) $2\frac{1}{2}$
13) $2\frac{3}{10}$
14) $\frac{4}{5}$
15) $\frac{1}{4}$

16) $\frac{7}{50}$ 18) $\frac{2}{25}$ 20) $2\frac{3}{5}$

17) $\frac{1}{5}$ 19) $\frac{9}{20}$ 21) $5\frac{1}{5}$

Factoring Numbers

1) 1, 2, 4, 17, 34, 68
2) 1, 2, 4, 7, 8, 14, 28, 56
3) 1, 2, 3, 4, 6, 8, 12, 24
4) 1, 2, 4, 5, 8, 10, 20, 40
5) 1, 2, 43, 86
6) 1, 2, 3, 6, 13, 26, 39, 78
7) 1, 2, 5, 10, 25, 50
8) 1, 2, 7, 14, 49, 98
9) 1, 3, 5, 9, 15, 45
10) 1, 2, 13, 26
11) 1, 2, 3, 6, 9, 18, 27, 54
12) 1, 2, 4, 7, 14, 28

13) 1, 5, 11, 55
14) 1, 5, 17, 85
15) 1, 2, 3, 4, 6, 8, 12, 16, 24, 48
16) $2 \times 5 \times 5$
17) 5×5
18) 3×23
19) 3×7
20) $3 \times 3 \times 5$
21) $2 \times 2 \times 17$
22) 2×13
23) 2×43
24) 3×31

Greatest Common Factor

1) 10
2) 2
3) 5
4) 4
5) 1
6) 3
7) 3

8) 2
9) 2
10) 1
11) 2
12) 3
13) 4
14) 10

15) 1
16) 5
17) 2
18) 5
19) 12
20) 2
21) 10

Least Common Multiple

1) 28
2) 15
3) 80
4) 68
5) 24
6) 24
7) 18

8) 30
9) 152
10) 63
11) 551
12) 42
13) 150
14) 8

15) 150
16) 108
17) 72
18) 72
19) 780
20) 120
21) 90

Divisibility Rules

1) 16

2) 10

3) 15

4) 28

5) 36

6) 18

7) 27

8) 70

9) 57

10) 102

11) 144

12) 75

<u>2</u> 3 <u>4</u> 5 6 7 <u>8</u> 9 10

<u>2</u> 3 4 <u>5</u> 6 7 8 9 <u>10</u>

2 <u>3</u> 4 <u>5</u> 6 7 8 9 10

<u>2</u> 3 <u>4</u> 5 6 <u>7</u> 8 9 10

<u>2</u> <u>3</u> <u>4</u> 5 <u>6</u> 7 8 <u>9</u> 10

<u>2</u> <u>3</u> 4 5 <u>6</u> 7 8 <u>9</u> 10

2 <u>3</u> 4 5 6 7 8 <u>9</u> 10

<u>2</u> 3 4 <u>5</u> 6 <u>7</u> 8 9 <u>10</u>

2 <u>3</u> 4 5 6 7 8 9 10

<u>2</u> <u>3</u> 4 5 <u>6</u> 7 8 9 10

<u>2</u> <u>3</u> <u>4</u> 5 <u>6</u> 7 <u>8</u> <u>9</u> 10

2 <u>3</u> 4 <u>5</u> 6 7 8 9 10

Chapter 3: Real Numbers and Integers

Topics that you'll learn in this chapter:

- ✓ Adding and Subtracting Integers
- ✓ Multiplying and Dividing Integers
- ✓ Ordering Integers and Numbers
- ✓ Arrange and Order, Comparing Integers
- ✓ Order of Operations
- ✓ Mixed Integer Computations
- ✓ Integers and Absolute Value

"Wherever there is number, there is beauty." –Proclus

Adding and Subtracting Integers

| *Helpful*

Hints | - | **Integers:** {... , –3, –2, –1, 0, 1, 2, 3, ...}
Includes: zero, counting numbers, and the negative of the counting numbers.

– Add a positive integer by moving to the right on the number line.

– Add a negative integer by moving to the left on the number line.

– Subtract an integer by adding its opposite. | **Example:**

$12 + 10 = 22$

$25 - 13 = 12$

$(-24) + 12 = -12$

$(-14) + (-12) = -26$

$14 - (-13) = 27$ |

🖎 **Find the sum.**

1) $(-12) + (-4)$

2) $5 + (-24)$

3) $(-14) + 23$

4) $(-8) + (39)$

5) $43 + (-12)$

6) $(-23) + (-4) + 3$

7) $4 + (-12) + (-10) + (-25)$

8) $19 + (-15) + 25 + 11$

9) $(-9) + (-12) + (32 - 14)$

10) $4 + (-30) + (45 - 34)$

🖎 **Find the difference.**

11) $(-14) - (-9) - (18)$

12) $(-9) - (-25)$

13) $(-12) - (8)$

14) $(28) - (-4)$

15) $(34) - (2)$

16) $(55) - (-5) + (-4)$

17) $(9) - (2) - (-5)$

18) $(2) - (4) - (-15)$

19) $(23) - (4) - (-34)$

20) $(-45) - (-87)$

Multiplying and Dividing Integers

Helpful *Hints*	(negative) × (negative) = positive (negative) ÷ (negative) = positive (negative) × (positive) = negative (negative) ÷ (positive) = negative (positive) × (positive) = positive	**Examples:** $3 \times 2 = 6$ $3 \times -3 = -9$ $-2 \times -2 = 4$ $10 \div 2 = 5$ $-4 \div 2 = -2$ $-12 \div -6 = 3$

✐ Find each product.

1) $(-8) \times (-2)$

2) 3×6

3) $(-4) \times 5 \times (-6)$

4) $2 \times (-6) \times (-6)$

5) $11 \times (-12)$

6) $10 \times (-5)$

7) 8×8

8) $(-8) \times (-9)$

9) $6 \times (-5) \times 3$

10) $6 \times (-1) \times 2$

✐ Find each quotient.

11) $18 \div 3$

12) $(-24) \div 4$

13) $(-63) \div (-9)$

14) $54 \div 9$

15) $20 \div (-2)$

16) $(-66) \div (-11)$

17) $64 \div 8$

18) $(-121) \div 11$

19) $72 \div 9$

20) $16 \div 4$

Ordering Integers and Numbers

Helpful *Hints*	To compare numbers, you can use number line! As you move from left to right on the number line, you find a bigger number!	**Example:** Order integers from least to greatest. $(-11, -13, 7, -2, 12)$ $-13 < -11 < -2 < 7 < 12$

✍ Order each set of integers from least to greatest.

1) $-15, -19, 20, -4, 1$ ___, ___, ___, ___, ___, ___

2) $6, -5, 4, -3, 2$ ___, ___, ___, ___, ___

3) $15, -42, 19, 0, -22$ ___, ___, ___, ___, ___, ___

4) $26, -91, 0, -13, 67, -55$ ___, ___, ___, ___, ___, ___

5) $-17, -71, 90, -25, -54, -39$ ___, ___, ___, ___, ___, ___

6) $98, 5, 46, 19, 77, 24$ ___, ___, ___, ___, ___, ___

✍ Order each set of integers from greatest to least.

7) $-2, 5, -3, 6, -4$ ___, ___, ___, ___, ___, ___

8) $-37, 7, -17, 27, 47$ ___, ___, ___, ___, ___, ___

9) $32, -27, 19, -17, 15$ ___, ___, ___, ___, ___, ___

10) $68, 81, 21, -18, 94, 72$ ___, ___, ___, ___, ___, ___

Arrange, Order, and Comparing Integers

Helpful *Hints*	When using a number line, numbers increase as you move to the right.	**Examples:** $5 < 7,$ $-5 < -2$ $-18 < -12$

🖎 *Arrange these integers in descending order.*

1) $21, 71, -18, -10, 82$ ___, ___, ___, ___, ___, ___

2) $15, 11, 20, 12, -9, -5$ ___, ___, ___, ___, ___, ___

3) $-5, 20, 15, 9, -11$ ___, ___, ___, ___, ___, ___

4) $19, 18, -9, -6, -11$ ___, ___, ___, ___, ___, ___

5) $56, -34, -12, -5, 32$ ___, ___, ___, ___, ___, ___

🖎 *Compare. Use >, =, <*

6) -8 ____ 12 11) -56 ____ -58

7) -10 ____ -16 12) 78 ____ 87

8) 43 ____ 34 13) -92 ____ -102

9) 15 ____ -16 14) -12 ____ -12

10) -354 ____ -345 15) -721 ____ -821

Order of Operations

			Example:
Helpful	-	Use "order of operations" rule when there are more than one math operation.	
Hints	-	PEMDAS (parentheses / exponents / multiply / divide / add / subtract)	$(12 + 4) \div (-4) = -4$

✎Evaluate each expression.

1) $(2 \times 2) + 5$

2) $24 - (3 \times 3)$

3) $(6 \times 4) + 8$

4) $25 - (4 \times 2)$

5) $(6 \times 5) + 3$

6) $64 - (2 \times 4)$

7) $25 + (1 \times 8)$

8) $(6 \times 7) + 7$

9) $48 \div (4 + 4)$

10) $(7 + 11) \div (-2)$

11) $9 + (2 \times 5) + 10$

12) $(5 + 8) \times \frac{3}{5} + 2$

13) $2 \times 7 - (\frac{10}{9 - 4})$

14) $(12 + 2 - 5) \times 7 - 1$

15) $(\frac{7}{5 - 1}) \times (2 + 6) \times 2$

16) $20 \div (4 - (10 - 8))$

17) $\frac{50}{4(5 - 4) - 3}$

18) $2 + (8 \times 2)$

Mixed Integer Computations

Helpful	**It worth remembering:**	**Example:**
	(negative) × (negative) = positive	
Hints	(negative) ÷ (negative) = positive	(−5) + 6 = 1
	(negative) × (positive) = negative	(−3) × (−2) = 6
	(negative) ÷ (positive) = negative	(9) ÷ (−3) = − 3
	(positive) × (positive) = positive	

✎ *Compute.*

1) $(-70) \div (-5)$

2) $(-14) \times 3$

3) $(-4) \times (-15)$

4) $(-65) \div 5$

5) $18 \times (-7)$

6) $(-12) \times (-2)$

7) $\dfrac{(-60)}{(-20)}$

8) $24 \div (-8)$

9) $22 \div (-11)$

10) $\dfrac{(-27)}{3}$

11) $4 \times (-4)$

12) $\dfrac{(-48)}{12}$

13) $(-14) \times (-2)$

14) $(-7) \times (7)$

15) $\dfrac{-30}{-6}$

16) $(-54) \div 6$

17) $(-60) \div (-5)$

18) $(-7) \times (-12)$

19) $(-14) \times 5$

20) $88 \div (-8)$

Integers and Absolute Value

Helpful *Hints*	To find an absolute value of a number, just find it's distance from 0!	**Example:** $\vert-6\vert = 6$ $\vert 6\vert = 6$ $\vert-12\vert = 12$ $\vert 12\vert = 12$

✎**Write absolute value of each number.**

1) -4

2) -7

3) -8

4) 4

5) 5

6) -10

7) 1

8) 6

9) 8

10) -2

11) -1

12) 10

13) 3

14) 7

15) -5

16) -3

17) -9

18) 2

19) 4

20) -6

21) 9

✎*Evaluate.*

22) $\vert-43\vert - \vert 12\vert + 10$

23) $76 + \vert-15 - 45\vert - \vert 3\vert$

24) $30 + \vert-62\vert - 46$

25) $\vert 32\vert - \vert-78\vert + 90$

26) $\vert-35 + 4\vert + 6 - 4$

27) $\vert-4\vert + \vert-11\vert$

28) $\vert-6 + 3 - 4\vert + \vert 7 + 7\vert$

29) $\vert-9\vert + \vert-19\vert - 5$

Answers of Worksheets – CHAPTER 3

Adding and Subtracting Integers

1) − 16
2) − 19
3) 9
4) 31
5) 31
6) − 24
7) − 43

8) 40
9) − 3
10) − 15
11) − 23
12) 16
13) − 20
14) 32

15) 32
16) 56
17) 12
18) 13
19) 53
20) 42

Multiplying and Dividing Integers

1) 16
2) 18
3) 120
4) 72
5) − 132
6) − 50
7) 64

8) 72
9) − 90
10) − 12
11) 6
12) − 6
13) 7
14) 6

15) − 10
16) 6
17) 8
18) − 11
19) 8
20) 4

Ordering Integers and Numbers

1) − 19, − 15, − 4, 1, 20
2) − 5, − 3, 2, 4, 6
3) − 42, − 22, 0, 15, 19
4) − 91, − 55, − 13, 0, 26, 67
5) − 71, − 54, − 39, − 25, − 17, 90

6) 5, 19, 24, 46, 77, 98
7) 6, 5, − 2, − 3, − 4
8) 47, 27, 7, − 17, − 37
9) 32, 19, 15, − 17, − 27
10) 94, 81, 72, 68, 21, − 18

Arrange and Order, Comparing Integers

1) 82, 71, 21, − 10, − 18

2) 20, 15, 12, 11, − 5, − 9

3) 20, 15, 9, − 5, −11

4) 19, 18, − 6, − 9, − 11

5) 56, 32, − 5, − 12, − 34

6) <

7) >

8) >

9) >

10) <

11) >

12) <

13) >

14) =

15) >

Order of Operations

1) 9

2) 15

3) 32

4) 17

5) 33

6) 56

7) 33

8) 49

9) 6

10) − 9

11) 29

12) 9.8

13) 12

14) 62

15) 28

16) 10

17) 50

18) 18

Mixed Integer Computations

1) 14

2) − 42

3) 60

4) − 13

5) − 126

6) 24

7) 3

8) − 3

9) − 2

10) − 9

11) − 16

12) − 4

13) 28

14) − 49

15) 5

16) − 9

17) 12

18) 84

19) − 70

20) − 11

Integers and Absolute Value

1) 4
2) 7
3) 8
4) 4
5) 5
6) 10
7) 1
8) 6
9) 8
10) 2

11) 1
12) 10
13) 3
14) 7
15) 5
16) 3
17) 9
18) 2
19) 4
20) 6

21) 9
22) 41
23) 133
24) 46
25) 44
26) 33
27) 15
28) 21
29) 23

Chapter 4: Proportions and Ratios

Topics that you'll learn in this chapter:

- ✓ Writing Ratios
- ✓ Simplifying Ratios
- ✓ Create a Proportion
- ✓ Similar Figures
- ✓ Simple Interest
- ✓ Ratio and Rates Word Problems

"Do not worry about your difficulties in mathematics. I can assure you mine are still greater." – Albert Einstein

Writing Ratios

Helpful *Hints*	– A ratio is a comparison of two numbers. Ratio can be written as a division.	**Example:** $3:5$, or $\dfrac{3}{5}$

✍ Express each ratio as a rate and unite rate.

1) 120 miles on 4 gallons of gas.

2) 24 dollars for 6 books.

3) 200 miles on 14 gallons of gas

4) 24 inches of snow in 8 hours

✍ Express each ratio as a fraction in the simplest form.

5) 3 feet out of 30 feet

6) 18 cakes out of 42 cakes

7) 16 dimes t0 24 dimes

8) 12 dimes out of 48 coins

9) 14 cups to 84 cups

10) 45 gallons to 65 gallons

11) 10 miles out of 40 miles

12) 22 blue cars out of 55 cars

13) 32 pennies to 300 pennies

14) 24 beetles out of 86 insects

Simplifying Ratios

Helpful *Hints*	– You can calculate equivalent ratios by multiplying or dividing both sides of the ratio by the same number.	**Examples:** 3 : 6 = 1 : 2 4 : 9 = 8 : 18

✎ *Reduce each ratio.*

1) 21 : 49

2) 20 : 40

3) 10: 50

4) 14: 18

5) 45: 27

6) 49: 21

7) 100: 10

8) 12 : 8

9) 35 : 45

10) 8: 20

11) 25: 35

12) 21 : 27

13) 52 : 82

14) 12: 36

15) 24 : 3

16) 15: 30

17) 3 : 36

18) 8 : 16

19) 6 : 100

20) 2 : 20

21) 10: 60

22) 14: 63

23) 68: 80

24) 8: 80

Create a Proportion

Helpful	– A proportion contains 2 equal fractions! A proportion simply means that two fractions are equal.	**Example:**
Hints		2, 4, 8, 16
		$\dfrac{2}{4} = \dfrac{8}{16}$

✎ Create proportion from the given set of numbers.

1) 1, 6, 2, 3

2) 12, 144, 1, 12

3) 16, 4, 8, 2

4) 9, 5, 27, 15

5) 7, 10, 60, 42

6) 8, 7, 24, 21

7) 10, 5, 8, 4

8) 3, 12, 8, 2

9) 2, 2, 1, 4

10) 3, 6, 7, 14

11) 2, 6, 5, 15

12) 7, 2, 14, 4

Similar Figures

| *Helpful* *Hints* | – Two or more figures are similar if the corresponding angles are equal, and the corresponding sides are in proportion. | **Example:**

3–4–5 triangle is similar to a

6–8–10 triangle |

✎ **Each pair of figures is similar. Find the missing side.**

1)

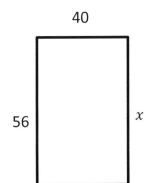
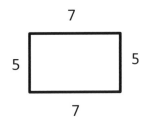

2)

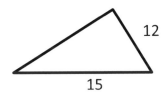
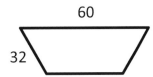

3)

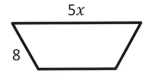

Simple Interest

Helpful *Hints*	**Simple Interest:** The charge for borrowing money or the return for lending it. Interest = principal x rate x time $$I = prt$$	**Example:** $450 at 7% for 8 years. $$I = prt$$ $$I = 450 \times 0.07 \times 8 = \$252 =$$

✎ **Use simple interest to find the ending balance.**

1) $1,300 at 5% for 6 years.

2) $5,400 at 7.5% for 6 months.

3) $25,600 at 9.2% for 5 years

4) $24,000 at 8.5% for 9 years.

5) $450 at 7% for 8 years.

6) $54,200 at 8% for 5 years.

7) $240 interest is earned on a principal of $1500 at a simple interest rate of 4% interest per year. For how many years was the principal invested?

8) A new car, valued at $28,000, depreciates at 9% per year from original price. Find the value of the car 3 years after purchase.

9) Sara puts $2,000 into an investment yielding 5% annual simple interest; she left the money in for five years. How much interest does Sara get at the end of those five years?

Ratio and Rates Word Problems

Helpful *Hints*	To solve a ratio or a rate word problem, create a proportion and use cross multiplication method!	**Example:** $\dfrac{x}{4} = \dfrac{8}{16}$ $16x = 4 \times 8$ $x = 2$

✎ **Solve.**

1) In a party, 10 soft drinks are required for every 12 guests. If there are 252 guests, how many soft drink is required?

2) In Jack's class, 18 of the students are tall and 10 are short. In Michael's class 54 students are tall and 30 students are short. Which class has a higher ratio of tall to short students?

3) Are these ratios equivalent?
   12 cards to 72 animals 11 marbles to 66 marbles

4) The price of 3 apples at the Quick Market is $1.44. The price of 5 of the same apples at Walmart is $2.50. Which place is the better buy?

5) The bakers at a Bakery can make 160 bagels in 4 hours. How many bagels can they bake in 16 hours? What is that rate per hour?

6) You can buy 5 cans of green beans at a supermarket for $3.40. How much does it cost to buy 35 cans of green beans?

Answers of Worksheets – Chapter 4

Writing Ratios

1) $\frac{120\ miles}{4\ gallons}$, 30 miles per gallon

2) $\frac{24\ dollars}{6\ books}$, 4.00 dollars per book

3) $\frac{200\ miles}{14\ gallons}$, 14.29 miles per gallon

4) $\frac{24"\ of\ snow}{8\ hours}$, 3 inches of snow per hour

5) $\frac{1}{10}$

6) $\frac{3}{7}$

7) $\frac{2}{3}$

8) $\frac{1}{4}$

9) $\frac{1}{6}$

10) $\frac{9}{13}$

11) $\frac{1}{4}$

12) $\frac{2}{5}$

13) $\frac{8}{75}$

14) $\frac{12}{43}$

Simplifying Ratios

1) 3 : 7
2) 1 : 2
3) 1 : 5
4) 7 : 9
5) 5 : 3
6) 7 : 3
7) 10 : 1
8) 3 : 2

9) 7 : 9
10) 2 : 5
11) 5 : 7
12) 7 : 9
13) 26 : 41
14) 1 : 3
15) 8 : 1
16) 1 : 2

17) 1 : 12
18) 1 : 2
19) 3 : 50
20) 1 : 10
21) 1 : 6
22) 2 : 9
23) 17 : 20
24) 1 : 10

Create a Proportion

1) 1 : 3 = 2 : 6

2) 12 : 144 = 1 : 12

3) 2 : 4 = 8 : 16

4) 5 : 15 = 9 : 27

5) 7 : 42, 10 : 60

6) 7 : 21 = 8 : 24

7) 8 : 10 = 4 : 5

8) 2 : 3 = 8 : 12

9) 4 : 2 = 2 : 1

10) 7 : 3 = 14 : 6   11) 5 : 2 = 15 : 6   12) 7 : 2 = 14 : 4

Similar Figures

1) 5   2) 3   3) 56

Simple Interest

1) $1,690.00   4) $42,360.00   7) 4 years

2) $5,602.50   5) $702.00   8) $20,440

3) $37,376.00   6) $75,880.00   9) $500

Ratio and Rates Word Problems

1) 210

2) The ratio for both class is equal to 9 to 5.

3) Yes! Both ratios are 1 to 6

4) The price at the Quick Market is a better buy.

5) 640, the rate is 40 per hour.

6) $23.80

Chapter 5: Percent

Topics that you'll learn in this chapter:

- ✓ Percentage Calculations
- ✓ Converting Between Percent, Fractions, and Decimals
- ✓ Percent Problems
- ✓ Markup, Discount, and Tax

"Do not worry about your difficulties in mathematics. I can assure you mine are still greater." –
Albert Einstein

Percentage Calculations

Helpful	-	Use the following formula to find part, whole, or percent:	**Example:**
Hints		$\text{part} = \dfrac{\text{percent}}{100} \times \text{whole}$	$\dfrac{20}{100} \times 100 = 20$

✎ **Calculate the percentages.**

1) 50% of 25

2) 80% of 15

3) 30% of 34

4) 70% of 45

5) 10% of 0

6) 80% of 22

7) 65% of 8

8) 78% of 54

9) 50% of 80

10) 20% of 10

11) 40% of 40

12) 90% of 0

13) 20% of 70

14) 55% of 60

15) 80% of 10

16) 20% of 880

17) 70% of 100

18) 80% of 90

✎ **Solve.**

19) 50 is what percentage of 75?

20) What percentage of 100 is 70

21) Find what percentage of 60 is 35.

22) 40 is what percentage of 80?

Converting Between Percent, Fractions, and Decimals

Helpful	— To a percent: Move the decimal point 2 places to the right and add the % symbol.	**Examples:**
Hints	— Divide by 100 to convert a number from percent to decimal.	$30\% = 0.3$
		$0.24 = 24\%$

✍ *Converting fractions to decimals.*

1) $\dfrac{50}{100}$

2) $\dfrac{38}{100}$

3) $\dfrac{15}{100}$

4) $\dfrac{80}{100}$

5) $\dfrac{7}{100}$

6) $\dfrac{35}{100}$

7) $\dfrac{90}{100}$

8) $\dfrac{20}{100}$

9) $\dfrac{7}{100}$

✍ *Write each decimal as a percent.*

10) 0.5

11) 0.9

12) 0.002

13) 0.524

14) 0.1

15) 0.03

16) 3.63

17) 0.008

18) 4.78

Percent Problems

Helpful	Base = Part ÷ Percent Part = Percent × Base Percent = Part ÷ Base	**Example:** 2 is 10% of 20.
Hints		2 ÷ 0.10 = 20
		2 = 0.10 × 20
		0.10 = 2 ÷ 20

✎ **Solve each problem.**

1) 51 is 340% of what?

2) 93% of what number is 97?

3) 27% of 142 is what number?

4) What percent of 125 is 29.3?

5) 60 is what percent of 126?

6) 67 is 67% of what?

7) 67 is 13% of what?

8) 41% of 78 is what?

9) 1 is what percent of 52.6?

10) What is 59% of 14 m?

11) What is 90% of 130 inches?

12) 16 inches is 35% of what?

13) 90% of 54.4 hours is what?

14) What percent of 33.5 is 21?

15) Liam scored 22 out of 30 marks in Algebra, 35 out of 40 marks in science and 89 out of 100 marks in mathematics. In which subject his percentage of marks in best?

16) Ella require 50% to pass. If she gets 280 marks and falls short by 20 marks, what were the maximum marks she could have got?

Markup, Discount, and Tax

Helpful	-	**Markup** = selling price – cost	**Example:**

Helpful

Hints

- **Markup** = selling price – cost
Markup rate = markup divided by the cost

- **Discount:**
Multiply the regular price by the rate of discount

Selling price =

original price – discount

- **Tax:**
To find tax, multiply the tax rate to the taxable amount (income, property value, etc.)

Example:

Original price of a microphone: $49.99, discount: 5%, tax: 5%

Selling price = 49.87

✎*Find the selling price of each item.*

1) Cost of a pen: $1.95, markup: 70%, discount: 40%, tax: 5%

2) Cost of a puppy: $349.99, markup: 41%, discount: 23%

3) Cost of a shirt: $14.95, markup: 25%, discount: 45%

4) Cost of an oil change: $21.95, markup: 95%

5) Cost of computer: $1,850.00, markup: 75%

Answers of Worksheets – Chapter 5

Percentage Calculations

1) 12.5	9) 40	17) 70
2) 12	10) 2	18) 72
3) 10.2	11) 16	19) 67%
4) 31.5	12) 0	20) 70%
5) 0	13) 14	21) 58%
6) 17.6	14) 33	22) 50%
7) 5.2	15) 8	
8) 42.12	16) 176	

Converting Between Percent, Fractions, and Decimals

1) 0.5	7) 0.9	13) 52.4%
2) 0.38	8) 0.2	14) 10%
3) 0.15	9) 0.07	15) 3%
4) 0.8	10) 50%	16) 363%
5) 0.07	11) 90%	17) 0.8%
6) 0.35	12) 0.2%	18) 478%

Percent Problems

1) 15	7) 515.4	13) 49 hours
2) 104.3	8) 31.98	14) 62.7%
3) 38.34	9) 1.9%	15) Mathematics
4) 23.44%	10) 8.3 m	16) 600
5) 47.6%	11) 117 inches	
6) 100	12) 45.7 inches	

Markup, Discount, and Tax

1) $2.09

2) $379.98

3) $10.28

4) $36.22

5) $3,237.50

Chapter 6: Algebraic Expressions

Topics that you'll learn in this chapter:

- ✓ Expressions and Variables
- ✓ Simplifying Variable Expressions
- ✓ Simplifying Polynomial Expressions
- ✓ Translate Phrases into an Algebraic Statement
- ✓ The Distributive Property
- ✓ Evaluating One Variable
- ✓ Evaluating Two Variables
- ✓ Combining like Terms

Without mathematics, there's nothing you can do. Everything around you is mathematics. Everything around you is numbers." – Shakuntala Devi

Expressions and Variables

Helpful *Hints*	A variable is a letter that represents unknown numbers. A variable can be used in the same manner as all other numbers:		
	Addition	$2 + a$	2 plus a
	Subtraction	$y - 3$	y minus 3
	Division	$\dfrac{4}{x}$	4 divided by x
	Multiplication	$5a$	5 times a

✎ **Simplify each expression.**

1) $x + 5x$,

   use $x = 5$

2) $8(-3x + 9) + 6$,

   use $x = 6$

3) $10x - 2x + 6 - 5$,

   use $x = 5$

4) $2x - 3x - 9$,

   use $x = 7$

5) $(-6)(-2x - 4y)$,

   use $x = 1$, $y = 3$

6) $8x + 2 + 4y$,

   use $x = 9$, $y = 2$

7) $(-6)(-8x - 9y)$,

   use $x = 5$, $y = 5$

8) $6x + 5y$,

   use $x = 7$, $y = 4$

✎ **Simplify each expression.**

9) $5(-4 + 2x)$

10) $-3 - 5x - 6x + 9$

11) $6x - 3x - 8 + 10$

12) $(-8)(6x - 4) + 12$

13) $9(7x + 4) + 6x$

14) $(-9)(-5x + 2)$

Simplifying Variable Expressions

Helpful	– Combine "like" terms. (values with same variable and same power)	**Example:**
Hints	– Use distributive property if necessary.	$2x + 2(1 - 5x) =$
		$2x + 2 - 10x = -8x + 2$
	Distributive Property:	
	$a(b + c) = ab + ac$	

✍ **Simplify each expression.**

1) $-2 - x^2 - 6x^2$

2) $3 + 10x^2 + 2$

3) $8x^2 + 6x + 7x^2$

4) $5x^2 - 12x^2 + 8x$

5) $2x^2 - 2x - x$

6) $(-6)(8x - 4)$

7) $4x + 6(2 - 5x)$

8) $10x + 8(10x - 6)$

9) $9(-2x - 6) - 5$

10) $3(x + 9)$

11) $7x + 3 - 3x$

12) $2.5x^2 \times (-8x)$

✍ **Simplify.**

13) $-2(4 - 6x) - 3x$, $x = 1$

14) $2x + 8x$, $x = 2$

15) $9 - 2x + 5x + 2$, $x = 5$

16) $5(3x + 7)$, $x = 3$

17) $2(3 - 2x) - 4$, $x = 6$

18) $5x + 3x - 8$, $x = 3$

19) $x - 7x$, $x = 8$

20) $5(-2 - 9x)$, $x = 4$

Simplifying Polynomial Expressions

Helpful	-	In mathematics, a polynomial is an expression consisting of variables and coefficients that involves only the operations of addition, subtraction, multiplication, and non–negative integer exponents of variables.	**Example:**

Helpful

Hints

In mathematics, a polynomial is an expression consisting of variables and coefficients that involves only the operations of addition, subtraction, multiplication, and non–negative integer exponents of variables.

$$P(x) = a_0 x^n + a_1 x^{n-1} + \ldots + a_{n-2} 2x^2 + a_{n-1} x + a_n$$

Example:

An example of a polynomial of a single indeterminate x is

$$x^2 - 4x + 7.$$

An example for three variables is

$$x^3 + 2xyz^2 - yz + 1$$

✍ *Simplify each polynomial.*

1) $4x^5 - 5x^6 + 15x^5 - 12x^6 + 3x^6$

2) $(-3x^5 + 12 - 4x) + (8x^4 + 5x + 5x^5)$

3) $10x^2 - 5x^4 + 14x^3 - 20x^4 + 15x^3 - 8x^4$

4) $-6x^2 + 5x^2 - 7x^3 + 12 + 22$

5) $12x^5 - 5x^3 + 8x^2 - 8x^5$ 11) $(12x^3 + 4x^4) - (2x^4 - 6x^3)$

6) $5x^3 + 1 + x^2 - 2x - 10x$ 12) $(12 + 3x^3) + (6x^3 + 6)$

7) $14x^2 - 6x^3 - 2x(4x^2 + 2x)$ 13) $(5x^2 - 3) + (2x^2 - 3x^3)$

8) $(4x^4 - 2x) - (4x - 2x^4)$ 14) $(23x^3 - 12x^2) - (2x^2 - 9x^3)$

9) $(3x^2 + 1) - (4 + 2x^2)$ 15) $(4x - 3x^3) - (3x^3 + 4x)$

10) $(2x + 2) - (7x + 6)$

Translate Phrases into an Algebraic Statement

Helpful	Translating key words and phrases into algebraic expressions:
	Addition: plus, more than, the sum of, etc.
Hints	**Subtraction:** minus, less than, decreased, etc.
	Multiplication: times, product, multiplied, etc.
	Division: quotient, divided, ratio, etc.
	Example:
	eight more than a number is 20
	$8 + x = 20$

✎ *Write an algebraic expression for each phrase.*

1) A number increased by forty–two.

2) The sum of fifteen and a number

3) The difference between fifty–six and a number.

4) The quotient of thirty and a number.

5) Twice a number decreased by 25.

6) Four times the sum of a number and − 12.

7) A number divided by − 20.

8) The quotient of 60 and the product of a number and − 5.

9) Ten subtracted from a number.

10) The difference of six and a number.

The Distributive Property

Helpful Hints	Distributive Property: $$a\,(b\,+\,c)\,=\,ab\,+\,ac$$	Example: $3\,(4\,+\,3x)$ $=\,12\,+\,9x$

✎ **Use the distributive property to simply each expression.**

1) $-\,(-\,2\,-\,5x)$

2) $(-\,6x\,+\,2)(-1)$

3) $(-\,5)\,(x\,-\,2)$

4) $-\,(7\,-\,3x)$

5) $8\,(8\,+\,2x)$

6) $2\,(12\,+\,2x)$

7) $(-\,6x\,+\,8)\,4$

8) $(3\,-\,6x)(-\,7)$

9) $(-\,12)\,(2x\,+\,1)$

10) $(8\,-\,2x)\,9$

11) $(-\,2x)\,(-\,1\,+\,9x)\,-\,4x\,(4\,+\,5x)$

12) $3\,(-\,5x\,-\,3)\,+\,4(6\,-\,3x)$

13) $(-\,2)(x\,+\,4)\,-\,(2\,+\,3x)$

14) $(-\,4)(3x\,-\,2)\,+\,6\,(x\,+\,1)$

15) $(-\,5)(4x\,-\,1)\,+\,4\,(x\,+\,2)$

16) $(-\,3)(x\,+\,4)\,-\,(2\,+\,3x)$

Evaluating One Variable

Helpful	– To evaluate one variable expression, find the variable and substitute a number for that variable.	**Example:**
Hints	– Perform the arithmetic operations.	$4x + 8, x = 6$
		$4(6) + 8 = 24 + 8 = 32$

✍ *Simplify each algebraic expression.*

1) $9 - x$, $x = 3$

2) $x + 2$, $x = 5$

3) $3x + 7$, $x = 6$

4) $x + (-5)$, $x = -2$

5) $3x + 6$, $x = 4$

6) $4x + 6$, $x = -1$

7) $10 + 2x - 6$, $x = 3$

8) $10 - 3x$, $x = 8$

9) $\dfrac{20}{x} - 3$, $x = 5$

10) $(-3) + \dfrac{x}{4} + 2x$, $x = 16$

11) $(-2) + \dfrac{x}{7}$, $x = 21$

12) $(-\dfrac{14}{x}) - 9 + 4x$, $x = 2$

13) $(-\dfrac{6}{x}) - 9 + 2x$, $x = 3$

14) $(-2) + \dfrac{x}{8}$, $x = 16$

Evaluating Two Variables

Helpful *Hints*	To evaluate an algebraic expression, substitute a number for each variable and perform the arithmetic operations.	**Example:** $2x + 4y - 3 + 2,$ $x = 5, y = 3$ $2(5) + 4(3) - 3 + 2$ $= 10 + 12 - 3 + 2$ $= 21$

✎ *Simplify each algebraic expression.*

1) $2x + 4y - 3 + 2,$

   $x = 5, y = 3$

2) $(-\dfrac{12}{x}) + 1 + 5y,$

   $x = 6, y = 8$

3) $(-4)(-2a - 2b),$

   $a = 5, b = 3$

4) $10 + 3x + 7 - 2y,$

   $x = 7, y = 6$

5) $9x + 2 - 4y,$

   $x = 7, y = 5$

6) $6 + 3(-2x - 3y),$

   $x = 9, y = 7$

7) $12x + y,$

   $x = 4, y = 8$

8) $x \times 4 \div y,$

   $x = 3, y = 2$

9) $2x + 14 + 4y,$

   $x = 6, y = 8$

10) $4a - (5 - b),$

   $a = 4, b = 6$

Combining like Terms

Helpful *Hints*	– Terms are separated by "+" and "−" signs. – Like terms are terms with same variables and same powers. – Be sure to use the "+" or "−" that is in front of the coefficient.	**Example:** $22x + 6 + 2x =$ $24x + 6$

✎ **Simplify each expression.**

1) $5 + 2x - 8$

2) $(-2x + 6)\,2$

3) $7 + 3x + 6x - 4$

4) $(-4) - (3)(5x + 8)$

5) $9x - 7x - 5$

6) $x - 12x$

7) $7\,(3x + 6) + 2x$

8) $(-11x) - 10x$

9) $3x - 12 - 5x$

10) $13 + 4x - 5$

11) $(-22x) + 8x$

12) $2\,(4 + 3x) - 7x$

13) $(-4x) - (6 - 14x)$

14) $5\,(6x - 1) + 12x$

15) $22x + 6 + 2x$

16) $(-13x) - 14x$

17) $(-6x) - 9 + 15x$

18) $(-6x) + 7x$

19) $(-5x) + 12 + 7x$

20) $(-3x) - 9 + 15x$

21) $20x - 19x$

Answers of Worksheets – Chapter 6

Expressions and Variables

1) 30
2) −66
3) 41
4) −16
5) 84

6) 82
7) 510
8) 62
9) $10x − 20$
10) $6 − 11x$

11) $3x + 2$
12) $44 − 48x$
13) $69x + 36$
14) $45x − 18$

Simplifying Variable Expressions

1) $−7x^2 − 2$
2) $10x^2 + 5$
3) $15x^2 + 6x$
4) $−7x^2 + 8x$
5) $2x^2 − 3x$
6) $−48x + 24$
7) $−26x + 12$

8) $90x − 48$
9) $−18x − 59$
10) $3x + 27$
11) $4x + 3$
12) $−20x^3$
13) 1
14) 20

15) 26
16) 80
17) $−22$
18) 16
19) $−48$
20) $−190$

Simplifying Polynomial Expressions

1) $−14x^6 + 19x^5$
2) $2x^5 + 8x^4 + x + 12$
3) $−33x^4 + 29x^3 + 10x^2$
4) $−7x^3 − x^2 + 34$
5) $4x^5 − 5x^3 + 8x^2$
6) $5x^3 + x^2 − 12x + 1$
7) $−14x^3 + 10x^2$
8) $6x^4 − 6x$

9) $x^2 − 3$
10) $−5x − 4$
11) $2x^4 + 18x^3$
12) $9x^3 + 18$
13) $−3x^3 + 7x^2 − 3$
14) $32x^3 − 14x^2$
15) $−6x^3$

Translate Phrases into an Algebraic Statement

1) $x + 42$
3) $56 − x$
4) $30/x$
5) $2x − 25$
8) $\dfrac{60}{−5x}$

2) $15 + x$
6) $4(x + (−12))$
7) $\dfrac{x}{−20}$
9) $x − 10$
10) $6 − x$

The Distributive Property

1) $5x + 2$
2) $6x - 2$
3) $-5x + 10$
4) $3x - 7$
5) $16x + 64$
6) $4x + 24$

7) $-24x + 32$
8) $42x - 21$
9) $-24x - 12$
10) $-18x + 72$
11) $-38x^2 - 14x$
12) $-27x + 15$

13) $-5x - 10$
14) $-6x + 14$
15) $-16x + 13$
16) $-6x - 14$

Evaluating One Variable

1) 6
2) 7
3) 25
4) -7
5) 18

6) 2
7) 10
8) -14
9) 1
10) 33

11) 1
12) -8
13) -5
14) 0

Evaluating Two Variables

1) 21
2) 39
3) 64
4) 26

5) 45
6) -111
7) 56
8) 6

9) 58
10) 17

Combining like Terms

1) $2x - 3$
2) $-4x + 12$
3) $9x + 3$
4) $-15x - 28$
5) $2x - 5$
6) $-11x$
7) $23x + 42$

8) $-21x$
9) $-2x - 12$
10) $4x + 8$
11) $-14x$
12) $-x + 8$
13) $10x - 6$
14) $42x - 5$

15) $24x + 6$
16) $-27x$
17) $9x - 9$
18) x
19) $2x + 12$
20) $12x - 9$
21) x

Chapter 7: Equations

Topics that you'll learn in this chapter:

- ✓ One– Step Equations
- ✓ Two– Step Equations
- ✓ Multi– Step Equations

"The study of mathematics, like the Nile, begins in minuteness but ends in magnificence."

✓ – Charles Caleb Colton

One–Step Equations

Helpful	-	The values of two expressions on both sides of an equation are equal. $$ax + b = c$$	**Example:**
			$-8x = 16$
Hints	-	You only need to perform one Math operation in order to solve the equation.	$x = -2$

✎ **Solve each equation.**

1) x + 3 = 17

2) 22 = (− 8) + x

3) 3x = (− 30)

4) (− 36) = (− 6x)

5) (− 6) = 4 + x

6) 2 + x = (− 2)

7) 20x = (− 220)

8) 18 = x + 5

9) (− 23) + x = (− 19)

10) 5x = (− 45)

11) x − 12 = (− 25)

12) x − 3 = (− 12)

13) (− 35) = x − 27

14) 8 = 2x

15) (− 6x) = 36

16) (− 55) = (− 5x)

17) x − 30 = 20

18) 8x = 32

19) 36 = (− 4x)

20) 4x = 68

21) 30x = 300

Two–Step Equations

Helpful	– You only need to perform two math operations (add, subtract, multiply, or divide) to solve the equation.	**Example:**
Hints	– Simplify using the inverse of addition or subtraction.	$-2(x-1)=42$ $(x-1)=-21$
	– Simplify further by using the inverse of multiplication or division.	$x=-20$

✎ **Solve each equation.**

1) $5(8+x)=20$

2) $(-7)(x-9)=42$

3) $(-12)(2x-3)=(-12)$

4) $6(1+x)=12$

5) $12(2x+4)=60$

6) $7(3x+2)=42$

7) $8(14+2x)=(-34)$

8) $(-15)(2x-4)=48$

9) $3(x+5)=12$

10) $\dfrac{3x-12}{6}=4$

11) $(-12)=\dfrac{x+15}{6}$

12) $110=(-5)(2x-6)$

13) $\dfrac{x}{8}-12=4$

14) $20=12+\dfrac{x}{4}$

15) $\dfrac{-24+x}{6}=(-12)$

16) $(-4)(5+2x)=(-100)$

17) $(-12x)+20=32$

18) $\dfrac{-2+6x}{4}=(-8)$

19) $\dfrac{x+6}{5}=(-5)$

20) $(-9)+\dfrac{x}{4}=(-15)$

Multi–Step Equations

Helpful *Hints*	– Combine "like" terms on one side. – Bring variables to one side by adding or subtracting. – Simplify using the inverse of addition or subtraction. – Simplify further by using the inverse of multiplication or division.	**Example:** $3x + 15 = -2x + 5$ Add 2x both sides $5x + 15 = +5$ Subtract 15 both sides $5x = -10$ Divide by 5 both sides $x = -2$

✎ *Solve each equation.*

1) $-(2 - 2x) = 10$

2) $-12 = -(2x + 8)$

3) $3x + 15 = (-2x) + 5$

4) $-28 = (-2x) - 12x$

5) $2(1 + 2x) + 2x = -118$

6) $3x - 18 = 22 + x - 3 + x$

7) $12 - 2x = (-32) - x + x$

8) $7 - 3x - 3x = 3 - 3x$

9) $6 + 10x + 3x = (-30) + 4x$

10) $(-3x) - 8(-1 + 5x) = 352$

11) $24 = (-4x) - 8 + 8$

12) $9 = 2x - 7 + 6x$

13) $6(1 + 6x) = 294$

14) $-10 = (-4x) - 6x$

15) $4x - 2 = (-7) + 5x$

16) $5x - 14 = 8x + 4$

17) $40 = -(4x - 8)$

18) $(-18) - 6x = 6(1 + 3x)$

19) $x - 5 = -2(6 + 3x)$

20) $6 = 1 - 2x + 5$

Answers of Worksheets – Chapter 7

One–Step Equations

1) 14
2) 30
3) − 10
4) 6
5) − 10
6) − 4
7) − 11

8) 13
9) 4
10) − 9
11) − 13
12) − 9
13) − 8
14) 4

15) − 6
16) 11
17) 50
18) 4
19) − 9
20) 17
21) 10

Two–Step Equations

1) − 4
2) 3
3) 2
4) 1
5) 0.5
6) $\frac{4}{3}$
7) $-\frac{73}{8}$

8) $\frac{2}{5}$
9) − 1
10) 12
11) − 87
12) − 8
13) 128
14) 32

15) − 48
16) 10
17) − 1
18) − 5
19) − 31
20) − 24

Multi–Step Equations

1) 6
2) 2
3) − 2
4) 2
5) − 20
6) 37
7) 22

8) $\frac{4}{3}$
9) − 4
10) − 8
11) − 6
12) 2
13) 8

14) 1
15) 5
16) − 6
17) − 8
18) − 1
19) − 1
20) 0

Chapter 8: Inequalities

Topics that you'll learn in this chapter:

- ✓ Graphing Single– Variable Inequalities
- ✓ One– Step Inequalities
- ✓ Two– Step Inequalities
- ✓ Multi– Step Inequalities

Without mathematics, there's nothing you can do. Everything around you is mathematics.
Everything around you is numbers." – Shakuntala Devi

Graphing Single–Variable Inequalities

Helpful	– Isolate the variable.
	– Find the value of the inequality on the number line.
Hints	– For less than or greater than draw open circle on the value of the variable.
	– If there is an equal sign too, then use filled circle.
	– Draw a line to the right direction.

✏️ *Draw a graph for each inequality.*

1) $-2 > x$

2) $5 \leq -x$

3) $x > 7$

4) $-x > 1.5$

One–Step Inequalities

Helpful	– Isolate the variable.	**Example:**
	– For dividing both sides by negative numbers, flip the direction of the inequality sign.	$x + 4 \geq 11$
Hints		$x \geq 7$

✎**Solve each inequality and graph it.**

1) $x + 9 \geq 11$

2) $x - 4 \leq 2$

3) $6x \geq 36$

4) $7 + x < 16$

5) $x + 8 \leq 1$

6) $3x > 12$

7) $3x < 24$

Two–Step Inequalities

Helpful *Hints*	– Isolate the variable. – For dividing both sides by negative numbers, flip the direction of the of the inequality sign. – Simplify using the inverse of addition or subtraction. – Simplify further by using the inverse of multiplication or division.	**Example:** $2x + 9 \geq 11$ $2x \geq 2$ $x \geq 1$

🖎 **Solve each inequality and graph it.**

1) $3x - 4 \leq 5$

2) $2x - 2 \leq 6$

3) $4x - 4 \leq 8$

4) $3x + 6 \geq 12$

5) $6x - 5 \geq 19$

6) $2x - 4 \leq 6$

7) $8x - 4 \leq 4$

8) $6x + 4 \leq 10$

9) $5x + 4 \leq 9$

10) $7x - 4 \leq 3$

11) $4x - 19 < 19$

12) $2x - 3 < 21$

13) $7 + 4x \geq 19$

14) $9 + 4x < 21$

15) $3 + 2x \geq 19$

16) $6 + 4x < 22$

Multi–Step Inequalities

Helpful *Hints*	– Isolate the variable.	**Example:**
	– Simplify using the inverse of addition or subtraction.	$\dfrac{7x + 1}{3} \geq 5$
	– Simplify further by using the inverse of multiplication or division.	$7x + 1 \geq 15$
		$7x \geq 14$
		$x \geq 7$

✎ **Solve each inequality.**

1) $\dfrac{9x}{7} - 7 < 2$

2) $\dfrac{4x + 8}{2} \leq 12$

3) $\dfrac{3x - 8}{7} > 1$

4) $-3\,(x - 7) > 21$

5) $4 + \dfrac{x}{3} < 7$

6) $\dfrac{2x + 6}{4} \leq 10$

Answers of Worksheets – Chapter 8

Graphing Single–Variable Inequalities

1) $-2 > x$

2) $x \leq -5$

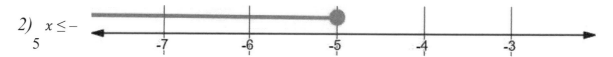

3) $x > 7$

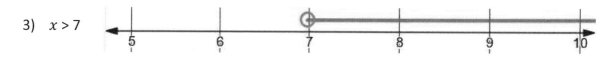

4) $-1.5 > x$

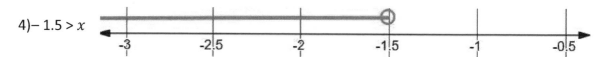

One–Step Inequalities

1)

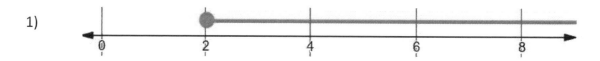

2)

3)

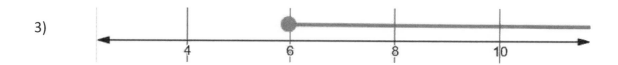

4)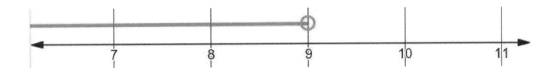

5)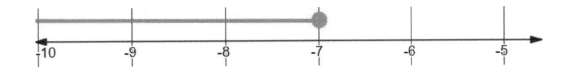

6)

7)

Two–Step inequalities

1) $x \leq 3$

2) $x \leq 4$

3) $x \leq 3$

4) $x \geq 2$

5) $x \geq 4$

6) $x \leq 5$

7) $x \leq 1$

8) $x \leq 1$

9) $x \leq 1$

10) $x \leq 1$

11) $x < 9.5$

12) $x < 12$

13) $x \geq 3$

14) $x < 3$

15) $x \geq 8$

16) $x < 4$

Multi–Step inequalities

1) $x < 7$

2) $x \leq 4$

3) $x > 5$

4) $x < 0$

5) $x < 9$

6) $x \leq 17$

Chapter 9: Linear Functions

Topics that you'll learn in this chapter:

- ✓ Finding Slope
- ✓ Graphing Lines Using Slope– Intercept Form
- ✓ Graphing Lines Using Standard Form
- ✓ Writing Linear Equations
- ✓ Graphing Linear Inequalities
- ✓ Finding Midpoint
- ✓ Finding Distance of Two Points

"Sometimes the questions are complicated and the answers are simple." – Dr. Seuss

Finding Slope

	Slope of a line:	Example:
Helpful *Hints*	$$\frac{y_2 - y_1}{x_2 - x_1} = \frac{rise}{run}$$	$(2, -10), (3, 6)$ slope = 16

✍ **Find the slope of the line through each pair of points.**

1) $(1, 1), (3, 5)$

2) $(4, -6), (-3, -8)$

3) $(7, -12), (5, 10)$

4) $(19, 3), (20, 3)$

5) $(15, 8), (-17, 9)$

6) $(6, -12), (15, -3)$

7) $(3, 1), (7, -5)$

8) $(3, -2), (-7, 8)$

9) $(15, -3), (-9, 5)$

10) $(-4, 7), (-6, -4)$

11) $(6, -8), (-11, -7)$

12) $(-6, 13), (17, -9)$

13) $(-10, -2), (-6, -5)$

14) $(4, 5), (-4, 10)$

15) $(-3, 1), (-17, 2)$

16) $(7, 0), (-13, -11)$

17) $(17, -13), (17, 8)$

18) $(12, 2), (-7, 5)$

Graphing Lines Using Slope–Intercept Form

Helpful	**Slope–intercept form:** given the slope m and the y–intercept b, then the equation of the line is:
Hints	$y = mx + b.$

Example:

$y = 8x - 3$

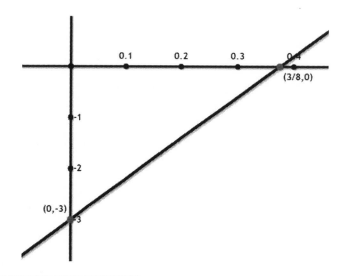

✎ **Sketch the graph of each line.**

1) $y = \frac{1}{2}x - 4$

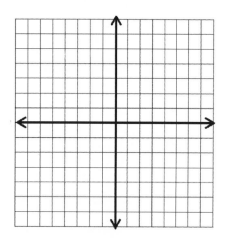

2) $y = 2x$

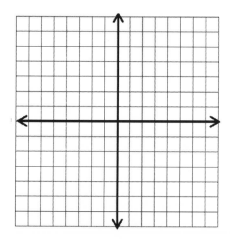

Graphing Lines Using Standard Form

Helpful *Hints*	– Find the –intercept of the line by putting zero for y.
	– Find the y–intercept of the line by putting zero for the x.
	– Connect these two points.

Example:

$x + 4y = 12$

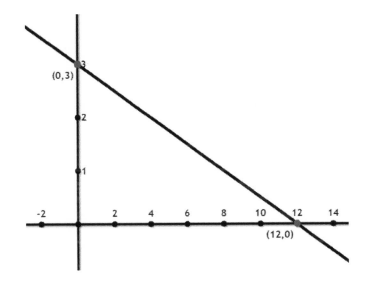

⬛️**Sketch the graph of each line.**

1) $2x - y = 4$

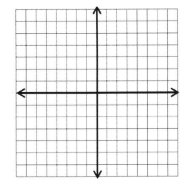

2) $x + y = 2$

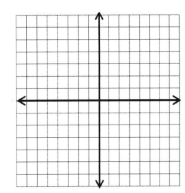

Writing Linear Equations

Helpful Hints	The equation of a line: $$y = mx + b$$ 1– Identify the slope. 2– Find the y–intercept. This can be done by substituting the slope and the coordinates of a point (x, y) on the line.	Example: through: $(-4, -2), (-3, 5)$ $y = 7x + 26$

✎ **Write the slope–intercept form of the equation of the line through the given points.**

1) through: $(-4, -2), (-3, 5)$

2) through: $(5, 4), (-4, 3)$

3) through: $(0, -2), (-5, 3)$

4) through: $(-1, 1), (-2, 6)$

5) through: $(0, 3), (-4, -1)$

6) through: $(0, 2), (1, -3)$

7) through: $(0, -5), (4, 3)$

8) through: $(-1, 4), (0, 4)$

9) through: $(2, -3), (3, -5)$

10) through: $(2, 5), (-1, -4)$

11) through: $(1, -3), (-3, 1)$

12) through: $(3, 3), (1, -5)$

13) through: $(4, 4), (3, -5)$

14) through: $(0, 3), (1, 1)$

15) through: $(5, 5), (2, -3)$

16) through: $(-2, -2), (2, -5)$

17) through: $(-3, -2), (1, -1)$

18) through: $(-2, 1), (6, 5)$

Graphing Linear Inequalities

Helpful	1– First, graph the "equals" line.
	2– Choose a testing point. (it can be any point on both sides of the line.)
Hints	3– Put the value of (x, y) of that point in the inequality. If that works, that part of the line is the solution. If the values don't work, then the other part of the line is the solution.

✍ *Sketch the graph of each linear inequality.*

1) $y < -4x + 2$

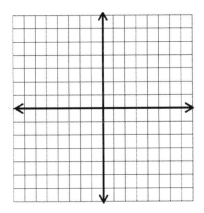

2) $2x + y < -4$

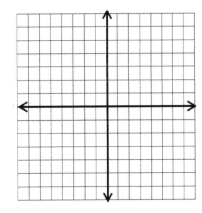

4) $x - 3y < -5$

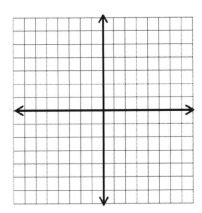

5) $6x - 2y \geq 8$

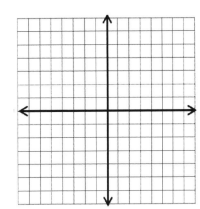

Finding Midpoint

Helpful	Midpoint of the segment AB:	**Example:**
Hints	$M\left(\dfrac{x_1+x_2}{2}, \dfrac{y_1+y_2}{2}\right)$	$(3, 9), (-1, 6)$ $M(1, 7.5)$

✎**Find the midpoint of the line segment with the given endpoints.**

1) $(2, -2), (3, -5)$

2) $(0, 2), (-2, -6)$

3) $(7, 4), (9, -1)$

4) $(4, -5), (0, 8)$

5) $(1, -2), (1, -6)$

6) $(-2, -3), (3, -6)$

7) $(7, 0), (-7, 5)$

8) $(-2, 6), (-3, -2)$

9) $(-1, 1), (5, -5)$

10) $(2.3, -1.3), (-2.2, -0.5)$

11) $(4.1, 6.32), (4, 5.6)$

12) $(2, -1), (-6, 0)$

13) $(-4, 4), (5, -1)$

14) $(-2, -3), (-6, 5)$

15) $\left(\dfrac{1}{2}, 1\right), (2, 4)$

16) $(-2, -2), (6, 5)$

Finding Distance of Two Points

Helpful	Distance from A to B:	**Example:**
Hints	$d = \sqrt{(x_1 - x_2)^2 + (y_1 - y_2)^2}$	$(-1, 2), (-1, -7)$ Distance = 9

✎ **Find the distance between each pair of points.**

1) $(2, -1), (1, -1)$

2) $(6, 4), (-1, 3)$

3) $(-8, -5), (-6, 1)$

4) $(-6, -10), (-2, -10)$

5) $(4, -6), (-3, 4)$

6) $(-6, -7), (-2, -8)$

7) $(5, 4), (8, 2)$

8) $(8, 4), (3, -7)$

9) $(1, 3), (5, 7)$

10) $(4, 2), (-7, 1)$

11) $(-3, -4), (-7, -2)$

12) $(-7, -2), (6, 9)$

13) $(10, 0), (0, 4)$

14) $(-3, 2), (5, 0)$

15) $(-5, 6), (8, -4)$

16) $(3, -5), (-8, -4)$

17) $(0, 8), (4, 10)$

18) $(6, 4), (-5, -1)$

Answers of Worksheets – Chapter 9

Finding Slope

1) 2

2) $\dfrac{2}{7}$

3) -11

4) 0

5) $-\dfrac{1}{32}$

6) 1

7) $-\dfrac{3}{2}$

8) -1

9) $-\dfrac{1}{3}$

10) $\dfrac{11}{2}$

11) $-\dfrac{1}{17}$

12) $-\dfrac{22}{23}$

13) $-\dfrac{3}{4}$

14) $-\dfrac{5}{8}$

15) $-\dfrac{1}{14}$

16) $\dfrac{11}{20}$

17) Undefined

18) $-\dfrac{3}{19}$

Graphing Lines Using Slope–Intercept Form

1)

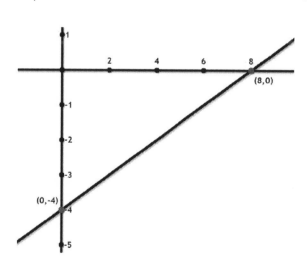

2)

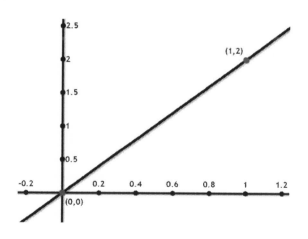

Graphing Lines Using Standard Form

1)

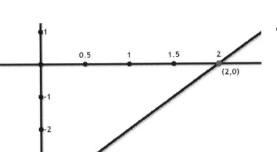

2)

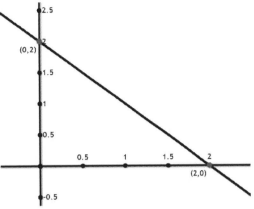

Writing Linear Equations

1) $y = 7x + 26$

2) $y = \frac{1}{9}x + \frac{31}{9}$

3) $y = -x - 2$

4) $y = -5x - 4$

5) $y = x + 3$

6) $y = -5x + 2$

7) $y = 2x - 5$

8) $y = 4$

9) $y = -2x + 1$

10) $y = 3x - 1$

11) $y = -x - 2$

12) $y = 4x - 9$

13) $y = 9x - 32$

14) $y = -2x + 3$

15) $y = \frac{8}{3}x - \frac{25}{3}$

16) $y = -\frac{3}{4}x - \frac{7}{2}$

17) $y = \frac{1}{4}x - \frac{5}{4}$

18) $y = -\frac{4}{3}x + \frac{19}{3}$

Graphing Linear Inequalities

1)

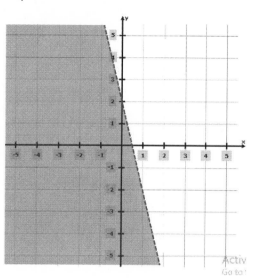

2)

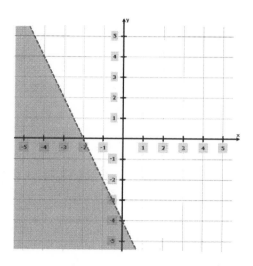

4)

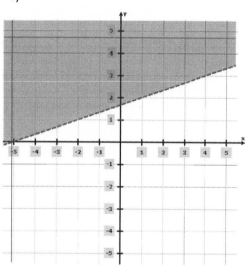

5)

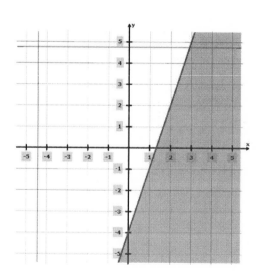

Finding Midpoint

1) (2.5, −3.5)

2) (−1, −2)

3) (8, 1.5)

4) (2, 1.5)

5) (1, −4)

6) (0.5, −4.5)

7) (0, 2.5)

8) (−2.5, 2)

9) (2, −2)

10) (0.05, −0.9)

11) (4.05, 5.96)

12) (−2, − 0.5)

13) $(\frac{1}{2}, 1\frac{1}{2})$

14) (−4, 1)

15) (1.25, 2.5)

16) $(2, \frac{3}{2})$

Finding Distance of Two Points

1) 1

2) 7.1

3) 6.32

4) 4

5) 12.21

6) 4.12

7) 3.61

8) 12.1

9) 5.66

10) 11.04

11) 4.47

12) 17.03

13) 10.77

14) 8.25

15) 16.4

16) 10.3

17) 4.47

18) 12.1

Chapter 10: Polynomials

Topics that you'll learn in this chapter:

✓ Classifying Polynomials

✓ Writing Polynomials in Standard Form

✓ Simplifying Polynomials

✓ Adding and Subtracting Polynomials

✓ Multiplying Monomials

✓ Multiplying and Dividing Monomials

✓ Multiplying a Polynomial and a Monomial

✓ Multiplying Binomials

✓ Factoring Trinomials

✓ Operations with Polynomials

Mathematics – the unshaken Foundation of Sciences, and the plentiful Fountain of Advantage to human

affairs. — Isaac Barrow

Classifying Polynomials

Helpful	Name	Degree	Example
	constant	0	4
Hints	linear	1	$2x$
	quadratic	2	$x^2 + 5x + 6$
	cubic	3	$x^3 - x^2 + 4x + 8$
	quartic	4	$x^4 + 3x^3 - x^2 + 2x + 6$
	quantic	5	$x^5 - 2x^4 + x^3 - x^2 + x + 10$

✎ **Name each polynomial by degree and number of terms.**

1) x

2) $-5x^4$

3) $7x - 4$

4) -6

5) $8x + 1$

6) $9x^2 - 8x^3$

7) $2x^5$

8) $10 + 8x$

9) $5x^2 - 6x$

10) $-7x^7 + 7x^4$

11) $-8x^4 + 5x^3 - 2x^2 - 8x$

12) $4x - 9x^2 + 4x^3 - 5x^4$

13) $4x^6 + 5x^5 + x^4$

14) $-4 - 2x^2 + 8x$

15) $9x^6 - 8$

16) $7x^5 + 10x^4 - 3x + 10x^7$

17) $4x^6 - 3x^2 - 8x^4$

18) $-5x^4 + 10x - 10$

Writing Polynomials in Standard Form

Helpful	A polynomial function $f(x)$ of degree n is of the form	**Example:**
Hints	$f(x) = a_nx^n + a_{n-1}x^{n-1} + \ldots + a_1x + a_0$ The first term is the one with the biggest power!	$2x^2 - 4x^3 - x =$ $-4x^3 + 2x^2 - x$

✍️ *Write each polynomial in standard form.*

1) $3x^2 - 5x^3$

2) $3 + 4x^3 - 3$

3) $2x^2 + 1x - 6x^3$

4) $9x - 7x$

5) $12 - 7x + 9x^4$

6) $5x^2 + 13x - 2x^3$

7) $-3 + 16x - 16x$

8) $3x(x + 4) - 2(x + 4)$

9) $(x + 5)(x - 2)$

10) $3x^2 + x + 12 - 5x^2 - 2x$

11) $12x^5 + 7x^3 - 3x^5 - 8x^3$

12) $3x(2x + 5 - 2x^2)$

13) $11x(x^5 + 2x^3)$

14) $(x + 6)(x + 3)$

15) $(x + 4)^2$

16) $(8x - 7)(3x + 2)$

17) $5x(3x^2 + 2x + 1)$

18) $7x(3 - x + 6x^3)$

Simplifying Polynomials

Helpful	1– Find "like" terms. (they have same variables with same power).	**Example:**
Hints	2– Add or Subtract "like" terms using PEMDAS operation.	$2x^5 - 3x^3 + 8x^2 - 2x^5 =$ $-3x^3 + 8x^2$

✎ *Simplify each expression.*

1) $11 - 4x^2 + 3x^2 - 7x^3 + 3$

2) $2x^5 - x^3 + 8x^2 - 2x^5$

3) $(-5)(x^6 + 10) - 8(14 - x^6)$

4) $4(2x^2 + 4x^2 - 3x^3) + 6x^3 + 17$

5) $11 - 6x^2 + 5x^2 - 12x^3 + 22$

6) $2x^2 - 2x + 3x^3 + 12x - 22x$

7) $(3x - 8)(3x - 4)$

8) $(12x + 2y)^2$

9) $(12x^3 + 28x^2 + 10x + 4) \div (x + 2)$

10) $(2x + 12x^2 - 2) \div (2x + 1)$

11) $(2x^3 - 1) + (3x^3 - 2x^3)$

12) $(x - 5)(x - 3)$

13) $(3x + 8)(3x - 8)$

14) $(8x^2 - 3x) - (5x - 5 - 8x^2)$

Adding and Subtracting Polynomials

Helpful *Hints*	Adding polynomials is just a matter of combining like terms, with some order of operations considerations thrown in. Be careful with the minus signs, and don't confuse addition and multiplication!	**Example:** $(3x^3 - 1) - (4x^3 + 2)$ $= -x^3 - 3$

✎ *Simplify each expression.*

1) $(2x^3 - 2) + (2x^3 + 2)$

6) $(4x^4 - 2x) - (6x - 2x^4)$

2) $(4x^3 + 5) - (7 - 2x^3)$

7) $(12x - 4x^3) - (8x^3 + 6x)$

3) $(4x^2 + 2x^3) - (2x^3 + 5)$

8) $(2x^3 - 8x^2) - (5x^2 - 3x^3)$

4) $(4x^2 - x) + (3x - 5x^2)$

9) $(2x^2 - 6) + (9x^2 - 4x^3)$

5) $(7x + 9) - (3x + 9)$

10) $(4x^3 + 3x^4) - (x^4 - 5x^3)$

11) $(-12x^4 + 10x^5 + 2x^3) + (14x^3 + 23x^5 + 8x^4)$

12) $(13x^2 - 6x^5 - 2x) - (-10x^2 - 11x^5 + 9x)$

13) $(35 + 9x^5 - 3x^2) + (8x^4 + 3x^5) - (27 - 5x^4)$

14) $(3x^5 - 2x^3 - 4x) + (4x + 10x^4 - 23) + (x^2 - x^3 + 12)$

Multiplying Monomials

Helpful *Hints*	A monomial is a polynomial with just one term, like $2x$ or $7y$.	**Example:** $2u^3 \times (-3u)$ $= -6u^4$

✎ **Simplify each expression.**

1) $2xy^2z \times 4z^2$

2) $4xy \times x^2y$

3) $4pq^3 \times (-2p^4q)$

4) $8s^4t^2 \times st^5$

5) $12p^3 \times (-3p^4)$

6) $-4p^2q^3r \times 6pq^2r^3$

7) $(-8a^4) \times (-12a^6b)$

8) $3u^4v^2 \times (-7u^2v^3)$

9) $4u^3 \times (-2u)$

10) $-6xy^2 \times 3x^2y$

11) $12y^2z^3 \times (-y^2z)$

12) $5a^2bc^2 \times 2abc^2$

Multiplying and Dividing Monomials

Helpful	- When you divide two monomials you need to divide their coefficients and then divide their variables.
Hints	- In case of exponents with the same base, you need to subtract their powers.

Example:

$(-3x^2)(8x^4y^{12}) = -24x^6y^{12}$

$\dfrac{36\,x^5y^7}{4\,x^4y^5} = 9xy^2$

✎ *Simplify.*

1) $(7x^4y^6)(4x^3y^4)$

2) $(15x^4)\,(3x^9)$

3) $(12x^2y^9)(7x^9y^{12})$

4) $\dfrac{80x^{12}y^9}{10x^6y^7}$

5) $\dfrac{95x^{18}y^7}{5x^9y^2}$

6) $\dfrac{200x^3y^8}{40x^3y^7}$

7) $\dfrac{-15x^{17}y^{13}}{3x^6y^9}$

8) $\dfrac{-64x^8y^{10}}{8x^3y^7}$

Multiplying a Polynomial and a Monomial

Helpful	– When multiplying monomials, use the product rule for exponents.	**Example:**
Hints	– When multiplying a monomial by a polynomial, use the distributive property.	$2x(8x-2) =$
	$a \times (b + c) = a \times b + a \times c$	$16x^2 - 4x$

✍ *Find each product.*

1) $5(3x - 6y)$

2) $9x(2x + 4y)$

3) $8x(7x - 4)$

4) $12x(3x + 9)$

5) $11x(2x - 11y)$

6) $2x(6x - 6y)$

7) $3x(2x^2 - 3x + 8)$

8) $13x(4x + 8y)$

9) $20(2x^2 - 8x - 5)$

10) $3x(3x - 2)$

11) $6x^3(3x^2 - 2x + 2)$

12) $8x^2(3x^2 - 5xy + 7y^2)$

13) $2x^2(3x^2 - 5x + 12)$

14) $2x^3(2x^2 + 5x - 4)$

15) $5x(6x^2 - 5xy + 2y^2)$

16) $9(x^2 + xy - 8y^2)$

Multiplying Binomials

Helpful	Use "FOIL". (First–Out–In–Last)	**Example:**
Hints	$(x + a)(x + b) = x^2 + (b + a)x + ab$	$(x + 2)(x - 3) =$ $x^2 - x - 6$

✎**Multiply.**

1) $(3x - 2)(4x + 2)$

2) $(2x - 5)(x + 7)$

3) $(x + 2)(x + 8)$

4) $(x^2 + 2)(x^2 - 2)$

5) $(x - 2)(x + 4)$

6) $(x - 8)(2x + 8)$

7) $(5x - 4)(3x + 3)$

8) $(x - 7)(x - 6)$

9) $(6x + 9)(4x + 9)$

10) $(2x - 6)(5x + 6)$

11) $(x - 7)(x + 7)$

12) $(x + 4)(4x - 8)$

13) $(6x - 4)(6x + 4)$

14) $(x - 7)(x + 2)$

15) $(x - 8)(x + 8)$

16) $(3x + 3)(3x - 4)$

17) $(x + 3)(x + 3)$

18) $(x + 4)(x + 6)$

Factoring Trinomials

Helpful Hints	"FOIL"	Example:
	$(x + a)(x + b) = x^2 + (b + a)x + ab$	$x^2 + 5x + 6 =$
	"Difference of Squares"	$(x + 2)(x + 3)$
	$a^2 - b^2 = (a + b)(a - b)$	
	$a^2 + 2ab + b^2 = (a + b)(a + b)$	
	$a^2 - 2ab + b^2 = (a - b)(a - b)$	
	"Reverse FOIL"	
	$x^2 + (b + a)x + ab = (x + a)(x + b)$	

✎Factor each trinomial.

1) $x^2 - 7x + 12$

2) $x^2 + 5x - 14$

3) $x^2 - 11x - 42$

4) $6x^2 + x - 12$

5) $x^2 - 17x + 30$

6) $x^2 + 8x + 15$

7) $3x^2 + 11x - 4$

8) $x^2 - 6x - 27$

9) $10x^2 + 33x - 7$

10) $x^2 + 24x + 144$

11) $49x^2 + 28xy + 4y^2$

12) $16x^2 - 40x + 25$

13) $x^2 - 10x + 25$

14) $25x^2 - 20x + 4$

15) $x^3 + 6x^2y^2 + 9xy^3$

16) $9x^2 + 24x + 16$

17) $x^2 - 8x + 16$

18) $x^2 + 121 + 22x$

Operations with Polynomials

Helpful	– When multiplying a monomial by a polynomial, use the distributive property.	**Example:**
Hints	a × (b + c) = a × b + a × c	$5(6x - 1) =$
		$30x - 5$

✍ **Find each product.**

1) $3x^2 (6x - 5)$

2) $5x^2 (7x - 2)$

3) $-3 (8x - 3)$

4) $6x^3 (-3x + 4)$

5) $9 (6x + 2)$

6) $8 (3x + 7)$

7) $5 (6x - 1)$

8) $-7x^4 (2x - 4)$

9) $8 (x^2 + 2x - 3)$

10) $4 (4x^2 - 2x + 1)$

11) $2 (3x^2 + 2x - 2)$

12) $8x (5x^2 + 3x + 8)$

13) $(9x + 1) (3x - 1)$

14) $(4x + 5) (6x - 5)$

15) $(7x + 3) (5x - 6)$

16) $(3x - 4) (3x + 8)$

Answers of Worksheets – Chapter 10

Classifying Polynomials

1) Linear monomial

2) Quartic monomial

3) Linear binomial

4) Constant monomial

5) Linear binomial

6) Cubic binomial

7) Quantic monomial

8) Linear binomial

9) Quadratic binomial

10) Seventh degree binomial

11) Quartic polynomial with four terms

12) Quartic polynomial with four terms

13) Sixth degree trinomial

14) Quadratic trinomial

15) Sixth degree binomial

16) Seventh degree polynomial with four terms

17) Sixth degree trinomial

18) Quartic trinomial

Writing Polynomials in Standard Form

1) $-5x^3 + 3x^2$

2) $4x^3$

3) $-6x^3 + 2x^2 + x$

4) $2x$

5) $9x^4 - 7x + 12$

6) $-2x^3 + 5x^2 + 13x$

7) -3

8) $3x^2 + 10x - 8$

9) $x^2 + 3x - 10$

10) $-2x^2 - x + 12$

11) $9x^5 - x^3$

12) $-6x^3 + 6x^2 + 15x$

13) $11x^6 + 22x^4$

14) $x^2 + 9x + 18$

15) $x^2 + 8x + 16$

16) $24x^2 - 5x - 14$

17) $15x^3 + 10x^2 + 5x$

18) $42x^4 - 7x^2 + 21x$

Simplifying Polynomials

1) $-7x^3 - x^2 + 14$

2) $-x^3 + 8x^2$

3) $3x^6 - 162$

4) $-6x^3 + 24x^2 + 17$

5) $-12x^3 - x^2 + 33$

6) $3x^3 + 2x^2 - 12x$

7) $9x^2 - 36x + 32$

8) $144x^2 + 48xy + 4y^2$

9) $12x^2 + 4x + 2$

10) $6x - 1$

11) $3x^3 - 1$

12) $x^2 - 8x + 15$

13) $9x^2 - 64$

14) $16x^2 - 8x + 5$

Adding and Subtracting Polynomials

1) $4x^3$

2) $6x^3 - 2$

3) $4x^2 - 5$

4) $-x^2 + 2x$

5) $4x$

6) $6x^4 - 8x$

7) $-12x^3 + 6x$

8) $5x^3 - 13x^2$

9) $-4x^3 + 11x^2 - 6$

10) $2x^4 + 9x^3$

11) $33x^5 - 4x^4 + 16x^3$

12) $5x^5 + 23x^2 - 11x$

13) $12x^5 + 13x^4 - 3x^2 + 8$

14) $3x^5 + 10x^4 - 3x^3 + x^2 - 11$

Multiplying Monomials

1) $8xy^2z^3$

2) $4x^3y^2$

3) $-8p^5q^4$

4) $8s^5t^7$

5) $-36p^7$

6) $-24p^3q^5r^4$

7) $96a^{10}b$

8) $-21u^6v^5$

9) $-8u^4$

10) $-18x^3y^3$

11) $-12y^4z^4$

12) $10a^3b^2c^4$

Multiplying and Dividing Monomials

1) $28x^7y^{10}$

2) $45x^{13}$

3) $84x^{11}y^{21}$

4) $8x^6y^2$

5) $19x^9y^5$

6) $5y$

7) $-5x^{11}y^4$

8) $-8x^5y^3$

Multiplying a Polynomial and a Monomial

1) $15x - 30y$

2) $18x^2 + 36xy$

3) $56x^2 - 32x$

4) $36x^2 + 108x$

5) $22x^2 - 121xy$

6) $12x^2 - 12xy$

7) $6x^3 - 9x^2 + 24x$

8) $52x^2 + 104xy$

9) $40x^2 - 160x - 100$

10) $9x^2 - 6x$

11) $18x^5 - 12x^4 + 12x^3$

12) $24x^4 - 40x^3y + 56y^2x^2$

13) $6x^4 - 10x^3 + 24x^2$

14) $4x^5 + 10x^4 - 8x^3$

15) $30x^3 - 25x^2y + 10xy^2$

16) $9x^2 + 9xy - 72y^2$

Multiplying Binomials

1) $12x^2 - 2x - 4$

2) $2x^2 + 9x - 35$

3) $x^2 + 10x + 16$

4) $x^4 - 4$

5) $x^2 + 2x - 8$

6) $2x^2 - 8x - 64$

7) $15x^2 + 3x - 12$

8) $x^2 - 13x + 42$

9) $24x^2 + 90x + 81$

10) $10x^2 - 18x - 36$

11) $x^2 - 49$

12) $4x^2 + 8x - 32$

13) $36x^2 - 16$

14) $x^2 - 5x - 14$

15) $x^2 - 64$

16) $9x^2 - 3x - 12$

17) $x^2 + 6x + 9$

18) $x^2 + 10x + 24$

Factoring Trinomials

1) $(x - 3)(x - 4)$

2) $(x - 2)(x + 7)$

3) $(x + 3)(x - 14)$

4) $(2x + 3)(3x - 4)$

5) $(x - 15)(x - 2)$

6) $(x + 3)(x + 5)$

7) $(3x - 1)(x + 4)$

8) $(x - 9)(x + 3)$

9) $(5x - 1)(2x + 7)$

10) $(x + 12)(x + 12)$

11) $(7x + 2y)(7x + 2y)$

12) $(4x - 5)(4x - 5)$

13) $(x - 5)(x - 5)$

14) $(5x - 2)(5x - 2)$

15) $x(x^2 + 6xy^2 + 9y^3)$

16) $(3x + 4)(3x + 4)$

17) $(x - 4)(x - 4)$

18) $(x + 11)(x + 11)$

Operations with Polynomials

1) $18x^3 - 15x^2$

2) $35x^3 - 10x^2$

3) $-24x + 9$

4) $-18x^4 + 24x^3$

5) $54x + 18$

6) $24x + 56$

7) $30x - 5$

8) $-14x^5 + 28x^4$

9) $8x^2 + 16x - 24$

10) $16x^2 - 8x + 4$

11) $6x^2 + 4x - 4$

12) $40x^3 + 24x^2 + 64x$

13) $27x^2 - 6x - 1$

14) $24x^2 + 10x - 25$

15) $35x^2 - 27x - 18$

16) $9x^2 + 12x - 32$

Chapter 11: Exponents and Radicals

Topics that you'll learn in this chapter:

✓ Multiplication Property of Exponents

✓ Division Property of Exponents

✓ Powers of Products and Quotients

✓ Zero and Negative Exponents

✓ Negative Exponents and Negative Bases

✓ Writing Scientific Notation

✓ Square Roots

Mathematics is no more computation than typing is literature.

– John Allen Paulos

Multiplication Property of Exponents

Helpful	Exponents rules	Example:
Hints	$x^a \cdot x^b = x^{a+b}$ $x^a/x^b = x^{a-b}$	$(x^2y)^3 = x^6y^3$
	$1/x^b = x^{-b}$ $(x^a)^b = x^{a.b}$	
	$(xy)^a = x^a \cdot y^a$	

✎ **Simplify.**

1) $4^2 \cdot 4^2$

2) $2 \cdot 2^2 \cdot 2^2$

3) $3^2 \cdot 3^2$

4) $3x^3 \cdot x$

5) $12x^4 \cdot 3x$

6) $6x \cdot 2x^2$

7) $5x^4 \cdot 5x^4$

8) $6x^2 \cdot 6x^3y^4$

9) $7x^2y^5 \cdot 9xy^3$

10) $7xy^4 \cdot 4x^3y^3$

11) $(2x^2)^2$

12) $3x^5y^3 \cdot 8x^2y^3$

13) $7x^3 \cdot 10y^3x^5 \cdot 8yx^3$

14) $(x^4)^3$

15) $(2x^2)^4$

16) $(x^2)^3$

17) $(6x)^2$

18) $3x^4y^5 \cdot 7x^2y^3$

Division Property of Exponents

Helpful	$\dfrac{x^a}{x^b} = x^{a-b}$, $x \neq 0$	Example:
Hints		$\dfrac{x^{12}}{x^5} = x^7$

✎ **Simplify.**

1) $\dfrac{5^5}{5}$

2) $\dfrac{3}{3^5}$

3) $\dfrac{2^2}{2^3}$

4) $\dfrac{2^4}{2^2}$

5) $\dfrac{x}{x^3}$

6) $\dfrac{3x^3}{9x^4}$

7) $\dfrac{2x^{-5}}{9x^{-2}}$

8) $\dfrac{21x^8}{7x^3}$

9) $\dfrac{7x^6}{4x^7}$

10) $\dfrac{6x^2}{4x^3}$

11) $\dfrac{5x}{10x^3}$

12) $\dfrac{3x^3}{2x^5}$

13) $\dfrac{12x^3}{14x^6}$

14) $\dfrac{12x^3}{9y^8}$

15) $\dfrac{25xy^4}{5x^6y^2}$

16) $\dfrac{2x^4}{7x}$

17) $\dfrac{16x^2y^8}{4x^3}$

18) $\dfrac{12x^4}{15x^7y^9}$

19) $\dfrac{12yx^4}{10yx^8}$

20) $\dfrac{16x^4y}{9x^8y^2}$

21) $\dfrac{5x^8}{20x^8}$

Powers of Products and Quotients

| *Helpful* *Hints* | For any nonzero numbers a and b and any integer x, $(ab)^x = a^x . b^x$. | **Example:** $(2x^2 . y^3)^2 =$ $4x^2 . y^6$ |

✎ **Simplify.**

1) $(2x^3)^4$

2) $(4xy^4)^2$

3) $(5x^4)^2$

4) $(11x^5)^2$

5) $(4x^2y^4)^4$

6) $(2x^4y^4)^3$

7) $(3x^2y^2)^2$

8) $(3x^4y^3)^4$

9) $(2x^6y^8)^2$

10) $(12x\ 3x)^3$

11) $(2x^9\ x^6)^3$

12) $(5x^{10}y^3)^3$

13) $(4x^3\ x^2)^2$

14) $(3x^3\ 5x)^2$

15) $(10x^{11}y^3)^2$

16) $(9x^7\ y^{\ 5})^2$

17) $(4x^4y^6)^5$

18) $(4x^4)^2$

19) $(3x\ 4y^3)^2$

20) $(9x^2y)^3$

21) $(12x^2y^5)^2$

Zero and Negative Exponents

Helpful *Hints*	A negative exponent simply means that the base is on the wrong side of the fraction line, so you need to flip the base to the other side. For instance, "x^{-2}" (pronounced as "ecks to the minus two") just means "x^2" but underneath, as in $\frac{1}{x^2}$	**Example:** $5^{-2} = \frac{1}{25}$

✎ Evaluate the following expressions.

1) 8^{-2}

2) 2^{-4}

3) 10^{-2}

4) 5^{-3}

5) 22^{-1}

6) 9^{-1}

7) 3^{-2}

8) 4^{-2}

9) 5^{-2}

10) 35^{-1}

11) 6^{-3}

12) 0^{15}

13) 10^{-9}

14) 3^{-4}

15) 5^{-2}

16) 2^{-3}

17) 3^{-3}

18) 8^{-1}

19) 7^{-3}

20) 6^{-2}

21) $\left(\frac{2}{3}\right)^{-2}$

22) $\left(\frac{1}{5}\right)^{-3}$

23) $\left(\frac{1}{2}\right)^{-8}$

24) $\left(\frac{2}{5}\right)^{-3}$

25) 10^{-3}

26) 1^{-10}

Negative Exponents and Negative Bases

Helpful *Hints*	– Make the power positive. A negative exponent is the reciprocal of that number with a positive exponent.	**Example:**
	– The parenthesis is important!	$2x^{-3} = \frac{2}{x^3}$
	– 5^{-2} is not the same as $(-5)^{-2}$	
	$-5^{-2} = -\frac{1}{5^2}$ and $(-5)^{-2} = +\frac{1}{5^2}$	

✎ **Simplify.**

1) -6^{-1}

2) $-4x^{-3}$

3) $-\frac{5x}{x^{-3}}$

4) $-\frac{a^{-3}}{b^{-2}}$

5) $-\frac{5}{x^{-3}}$

6) $\frac{7b}{-9c^{-4}}$

7) $-\frac{5n^{-2}}{10p^{-3}}$

8) $\frac{4ab^{-2}}{-3c^{-2}}$

9) $-12x^2y^{-3}$

10) $\left(-\frac{1}{3}\right)^{-2}$

11) $\left(-\frac{3}{4}\right)^{-2}$

12) $\left(\frac{3a}{2c}\right)^{-2}$

13) $\left(-\frac{5x}{3yz}\right)^{-3}$

14) $-\frac{2x}{a^{-4}}$

Writing Scientific Notation

Helpful	– It is used to write very big or very small numbers in decimal form.

Hints – In scientific notation all numbers are written in the form of:

$$m \times 10^n$$

Decimal notation	Scientific notation
5	5×10^0
−25,000	$−2.5 \times 10^4$
0.5	5×10^{-1}
2,122.456	$2,122456 \times 10^3$

✍ *Write each number in scientific notation.*

1) 91×10^3

2) 60

3) 2000000

4) 0.0000006

5) 354000

6) 0.000325

7) 2.5

8) 0.00023

9) 56000000

10) 2000000

11) 78000000

12) 0.0000022

13) 0.00012

14) 0.004

15) 78

16) 1600

17) 1450

18) 130000

19) 60

20) 0.113

21) 0.02

Square Roots

Helpful	— A square root of x is a number r whose square is: $r^2 = x$	Example:
Hints	r is a square root of x.	$\sqrt{4} = 2$

🖎 Find the value each square root.

1) $\sqrt{1}$

2) $\sqrt{4}$

3) $\sqrt{9}$

4) $\sqrt{25}$

5) $\sqrt{16}$

6) $\sqrt{49}$

7) $\sqrt{36}$

8) $\sqrt{0}$

9) $\sqrt{64}$

10) $\sqrt{81}$

11) $\sqrt{121}$

12) $\sqrt{225}$

13) $\sqrt{144}$

14) $\sqrt{100}$

15) $\sqrt{256}$

16) $\sqrt{289}$

17) $\sqrt{324}$

18) $\sqrt{400}$

19) $\sqrt{900}$

20) $\sqrt{529}$

21) $\sqrt{90}$

Answers of Worksheets – Chapter 11

Multiplication Property of Exponents

1) 4^4
2) 2^5
3) 3^4
4) $3x^4$
5) $36x^5$
6) $12x^3$

7) $25x^8$
8) $36x^5y^4$
9) $63x^3y^8$
10) $28x^4y^7$
11) $4x^4$
12) $24x^7y^6$

13) $560x^{11}y^4$
14) x^{12}
15) $16x^8$
16) x^6
17) $36x^2$
18) $21x^6y^8$

Division Property of Exponents

1) 5^4
2) $\frac{1}{3^4}$
3) $\frac{1}{2}$
4) 2^2
5) $\frac{1}{x^2}$
6) $\frac{1}{3x}$
7) $\frac{2}{9x^3}$
8) $3x^5$

9) $\frac{7}{4x}$
10) $\frac{3}{2x}$
11) $\frac{1}{2x^2}$
12) $\frac{3}{2x^2}$
13) $\frac{6}{7x^3}$
14) $\frac{4x^3}{3y^8}$
15) $\frac{5y^2}{x^5}$

16) $\frac{2x^3}{7}$
17) $\frac{4y^8}{x}$
18) $\frac{4}{5x^3y^9}$
19) $\frac{6}{5x^4}$
20) $\frac{16}{9x^4y}$
21) $\frac{1}{4}$

Powers of Products and Quotients

1) $16x^{12}$
2) $16x^2y^8$
3) $25x^8$
4) $121x^{10}$
5) $256x^8y^{16}$
6) $8x^{12}y^{12}$

7) $9x^4y^4$
8) $81x^{16}y^{12}$
9) $4x^{12}y^{16}$
10) $46,656x^6$
11) $8x^{45}$
12) $125x^{30}y^9$

13) $16x^{10}$
14) $225x^8$
15) $100x^{22}y^6$
16) $81x^{14}y^{10}$
17) $1,024x^{20}y^{30}$
18) $16x^8$

19) $144x^2y^6$

20) $729x^6y^3$

21) $144x^4y^{10}$

Zero and Negative Exponents

1) $\frac{1}{64}$

2) $\frac{1}{16}$

3) $\frac{1}{100}$

4) $\frac{1}{125}$

5) $\frac{1}{22}$

6) $\frac{1}{9}$

7) $\frac{1}{9}$

8) $\frac{1}{16}$

9) $\frac{1}{25}$

10) $\frac{1}{35}$

11) $\frac{1}{216}$

12) 0

13) $\frac{1}{1000000000}$

14) $\frac{1}{81}$

15) $\frac{1}{25}$

16) $\frac{1}{8}$

17) $\frac{1}{27}$

18) $\frac{1}{8}$

19) $\frac{1}{343}$

20) $\frac{1}{36}$

21) $\frac{9}{4}$

22) 125

23) 256

24) $\frac{125}{8}$

25) $\frac{1}{1000}$

26) 1

Negative Exponents and Negative Bases

1) $-\frac{1}{6}$

2) $-\frac{4}{x^3}$

3) $-5x^4$

4) $-\frac{b^2}{a^3}$

5) $-5x^3$

6) $-\frac{7bc^4}{9}$

7) $-\frac{p^3}{2n^2}$

8) $-\frac{4ac^2}{3b^2}$

9) $-\frac{12x^2}{y^3}$

10) 9

11) $\frac{16}{9}$

12) $\frac{4c^2}{9a^2}$

13) $-\frac{27y^3z^3}{125x^3}$

14) $-2xa^4$

Writing Scientific Notation

1) 9.1×10^4

2) 6×10^1

3) 2×10^6

4) 6×10^{-7}

5) 3.54×10^5

6) 3.25×10^{-4}

7) 2.5×10^0

8) 2.3×10^{-4}

9) 5.6×10^7

10) 2×10^6

11) 7.8×10^7

12) 2.2×10^{-6}

13) 1.2×10^{-4}

14) 4×10^{-3}

15) 7.8×10^{1}

16) 1.6×10^{3}

17) 1.45×10^{3}

18) 1.3×10^{5}

19) 6×10^{1}

20) 1.13×10^{-1}

21) 2×10^{-2}

Square Roots

1) 1

2) 2

3) 3

4) 5

5) 4

6) 7

7) 6

8) 0

9) 8

10) 9

11) 11

12) 15

13) 12

14) 10

15) 16

16) 17

17) 18

18) 20

19) 30

20) 23

21) $3\sqrt{10}$

Chapter 12: Statistics

Topics that you'll learn in this chapter:

✓ Mean, Median, Mode, and Range of the Given Data

✓ The Pie Graph or Circle Graph

✓ Probability

Mean, Median, Mode, and Range of the Given Data

Helpful			**Example:**
	-	Mean: $\dfrac{\text{sum of the data}}{\text{of data entires}}$	
Hints	-	Mode: value in the list that appears most often	22, 16, 12, 9, 7, 6, 4, 6
	-	Range: largest value – smallest value	Mean = 10.25
			Mod = 6
			Range = 18

✎ *Find Mean, Median, Mode, and Range of the Given Data.*

1) 7, 2, 5, 1, 1, 2

2) 2, 2, 2, 3, 6, 3, 7, 4

3) 9, 4, 3, 1, 7, 9, 4, 6, 4

4) 8, 4, 2, 4, 3, 2, 4, 5

5) 8, 5, 7, 5, 7, 9, 8

6) 5, 1, 4, 4, 9, 2, 9, 2, 5, 1

7) 4, 1, 5, 9, 7, 7, 5, 4, 3, 5

8) 7, 5, 4, 9, 6, 7, 7, 5, 2

9) 2, 5, 5, 6, 2, 4, 7, 6, 4, 9

10) 10, 5, 2, 5, 4, 5, 8, 10

11) 5, 1, 5, 2, 2

12) 2, 3, 5, 9, 6

The Pie Graph or Circle Graph

Helpful *Hints*	A Pie Chart is a circle chart divided into sectors, each sector represents the relative size of each value.

The circle graph below shows all Jason's expenses for last month. Jason spent $300 on his bills last month.

Answer following questions based on the Pie graph.

Jason's monthly expenses

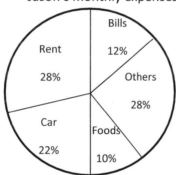

1- How much did Jason spend on his car last month?

2- How much did Jason spend for foods last month?

3- How much did Jason spend on his rent last month?

4- What fraction is Jason's expenses for his bills and Car out of his total expenses last month?

5- How much was Jason's senses last month?

Probability of Simple Events

Helpful *Hints*	- Probability is the likelihood of something happening in the future. It is expressed as a number between zero (can never happen) to 1 (will always happen). - Probability can be expressed as a fraction, a decimal, or a percent.

Example:

Probability of a flipped coins turns up 'heads'

Is $0.5 = \dfrac{1}{2}$

Solve.

1) A number is chosen at random from 1 to 10. Find the probability of selecting a 4 or smaller.

2) There are 135 blue balls and 15 red balls in a basket. What is the probability of randomly choosing a red ball from the basket?

3) A number is chosen at random from 1 to 10. Find the probability of selecting of 4 and factors of 6.

4) What is the probability of choosing a Hearts in a deck of cards? (A deck of cards contains 52 cards)

5) A number is chosen at random from 1 to 50. Find the probability of selecting prime numbers.

6) A number is chosen at random from 1 to 25. Find the probability of not selecting a composite number.

Answers of Worksheets – Chapter 12

Mean, Median, Mode, and Range of the Given Data

1) mean: 3, median: 2, mode: 1, 2, range: 6
2) mean: 3.625, median: 3, mode: 2, range: 5
3) mean: 5.22, median: 4, mode: 4, range: 8
4) mean: 4, median: 4, mode: 4, range: 6
5) mean: 7, median: 7, mode: 5, 7, 8, range: 4
6) mean: 4.2, median: 4, mode: 1,2,4,5,9, range: 8
7) mean: 5, median: 5, mode: 5, range: 8
8) mean: 5.78, median: 6, mode: 7, range: 7
9) mean: 5, median: 5, mode: 2, 4, 5, 6, range: 7
10) mean: 6.125, median: 5, mode: 5, range: 8
11) mean: 3, median: 2, mode: 2, 5, range: 4
12) mean: 5, median: 5, mode: none, range: 7

The Pie Graph or Circle Graph

1) $550
2) $250
3) $700
4) $\frac{17}{50}$
5) $2500

Probability of simple events

1) $\frac{2}{5}$

2) $\frac{1}{10}$

3) $\frac{1}{5}$

4) $\frac{1}{4}$

5) $\frac{3}{10}$

6) $\frac{2}{5}$

Chapter 13: Geometry

Topics that you'll learn in this chapter:

✓ The Pythagorean Theorem

✓ Area of Triangles

✓ Perimeter of Polygons

✓ Area and Circumference of Circles

✓ Area of Squares, Rectangles, and Parallelograms

✓ Area of Trapezoids

Mathematics is, as it were, a sensuous logic, and relates to philosophy as do the arts, music, and plastic

art to poetry. — K. Shegel

The Pythagorean Theorem

Helpful	– In any right triangle:	**Example:**

Hints

$a^2 + b^2 = c^2$

Missing side = 6

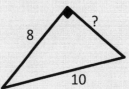

8 ?
10

🖋️ **Do the following lengths form a right triangle?**

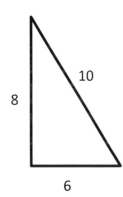

10
8
6

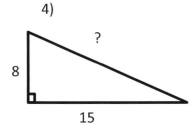

3 4
5

5 12
13

🖋️ **Find each missing length to the nearest tenth.**

4)

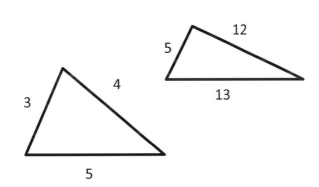

?
8
15

5)

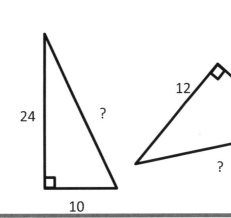

24 ?
10

6)

12 5
?

Area of Triangles

Helpful

Hints

$\text{Area} = \dfrac{1}{2}\ (base\ \times height)$

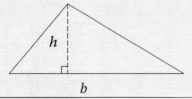

🖊 **Find the area of each.**

1)

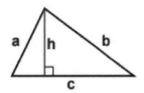

c = 9 mi

h = 3.7 mi

2)

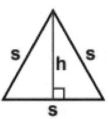

s = 14 m

h = 12.2 m

3)

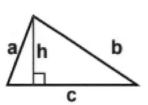

a = 5 m

b = 11 m

c = 14 m

h = 4 m

4)

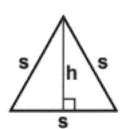

s = 10 m

h = 8.6 m

Perimeter of Polygons

Helpful

Hints

Perimeter of a square = 4s

 s

Perimeter of a rectangle

= 2(*l* + *w*)

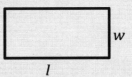 w

l

Perimeter of trapezoid

= a + b + c + d

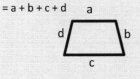

a

d   b

c

Perimeter of Pentagon = 6a

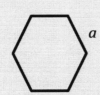 a

Perimeter of a parallelogram = 2(l + w)

l

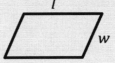 w

Example:

P = 18

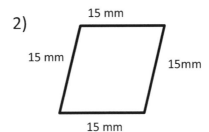

3 m

3 m   3 m

✎**Find the perimeter of each shape.**

1)

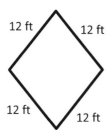

5 m

5 m   5 m

2)

15 mm

15 mm   15mm

15 mm

3)

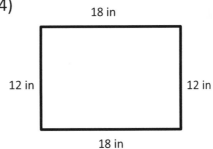

12 ft   12 ft

12 ft   12 ft

4)

18 in

12 in   12 in

18 in

Area and Circumference of Circles

Helpful	Area = πr²
	Circumference = 2πr
Hints	

Area = πr²

Circumference = 2πr

Example:

If the radius of a circle is 3, then:

Area = 28.27

Circumference = 18.85

✎ *Find the area and circumference of each.* (π = 3.14)

1)

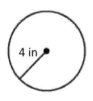

4 in

2)

18 cm

3)

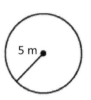

5 m

4)

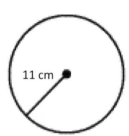

11 cm

5)

8 km

6)

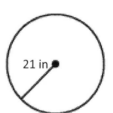

21 in

Area of Squares, Rectangles, and Parallelograms

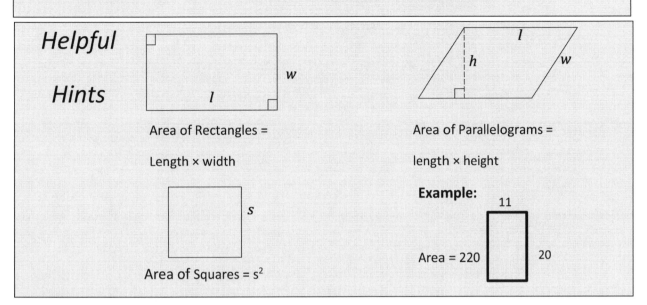

Helpful

Hints

Area of Rectangles =

Length × width

Area of Squares = s²

Area of Parallelograms =

length × height

Example:

11

Area = 220   20

📝 *Find the area of each.*

1)

22 yd

32.3 yd   32.3 yd

22 yd

2)

27mi

27 mi   27 mi

27 mi

3)

14.9 ft

15.1 ft

7 ft

15.1 ft

14.9 ft

4)

4 in

5.9 in

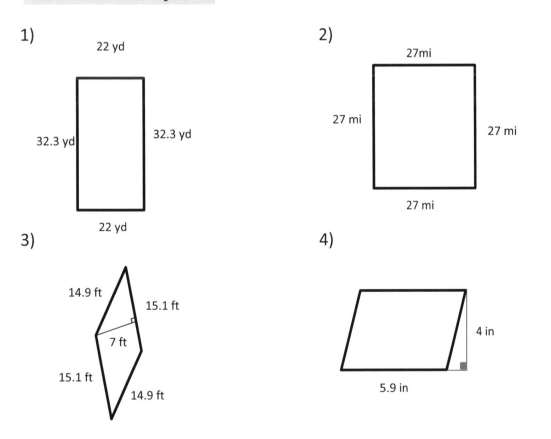

Area of Trapezoids

Helpful $A = \frac{1}{2}h(b_1 + b_2)$ **Example:**

Hints

A = 252 cm²

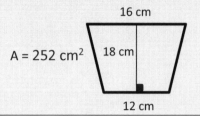

16 cm

18 cm

12 cm

✍ **Calculate the area for each trapezoid.**

1)

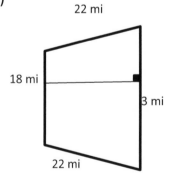

9 cm

6 cm

12 cm

2)

14 m

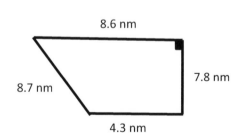

10 m

18 m

3)

22 mi

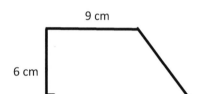

18 mi

3 mi

22 mi

4)

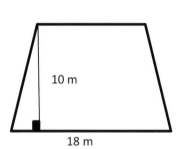

8.6 nm

8.7 nm

7.8 nm

4.3 nm

Answers of Worksheets – Chapter 13

The Pythagorean Theorem

1) yes

2) yes

3) yes

4) 17

5) 26

6) 13

Area of Triangles

1) 16.65 mi^2

2) 85.4 m^2

3) 28 m^2

4) 43 m^2

Perimeter of Polygons

1) 30 m

2) 60 mm

3) 48 ft

4) 60 in

Area and Circumference of Circles

1) Area: 50.24 in^2, Circumference: 25.12 in

2) Area: 1,017.36 cm^2, Circumference: 113.04 cm

3) Area: 78.5m^2, Circumference: 31.4 m

4) Area: 379.94 cm^2, Circumference: 69.08 cm

5) Area: 200.96 km^2, Circumference: 50.2 km

6) Area: 1,384.74 km^2, Circumference: 131.88 km

Area of Squares, Rectangles, and Parallelograms

1) 710.6 yd^2

2) 729 mi^2

3) 105.7 ft^2

4) 23.6 in^2

Area of Trapezoids

1) 63 cm^2

2) 160 m^2

3) 410 mi^2

4) 50.31 nm^2

Chapter 14: Solid Figures

Topics that you'll learn in this chapter:

✓ Volume of Cubes

✓ Volume of Rectangle Prisms

✓ Surface Area of Cubes

✓ Surface Area of Rectangle Prisms

✓ Volume of a Cylinder

✓ Surface Area of a Cylinder

Mathematics is a great motivator for all humans. Because its career starts with zero and it never end

(infinity)

Volume of Cubes

Helpful	– Volume is the measure of the amount of space inside of a solid figure, like a cube, ball, cylinder or pyramid.
Hints	– Volume of a cube = (one side)^3
	– Volume of a rectangle prism: Length × Width × Height

✎ *Find the volume of each.*

1)
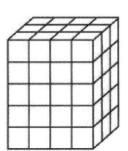

2)
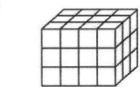

3)

4)

5)

6)
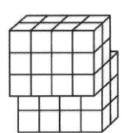

Volume of Rectangle Prisms

Helpful	Volume of rectangle prism	**Example:**
Hints	length × width × height	$10 \times 5 \times 8 = 400$

✍️ **Find the volume of each of the rectangular prisms.**

1)

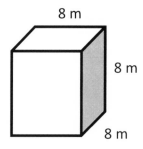

2)

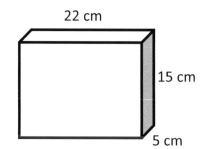

3)

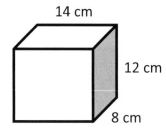

4)

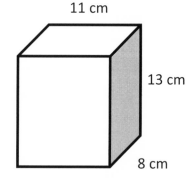

Surface Area of Cubes

Helpful

Hints

Surface Area of a cube =

6 × (one side of the cube)^2

Example:

$6 × 4^2 = 96m^3$

4 m

4 m

4 m

🖎 *Find the surface of each cube.*

1)

6 mm

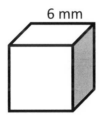

2)

9 mm

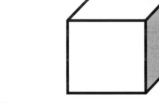

3)

10 cm

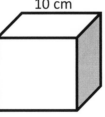

4)

8 m

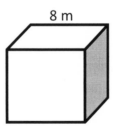

5)

7.5 in

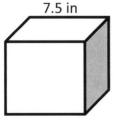

6)

11.3 ft

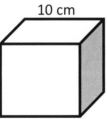

Surface Area of a Rectangle Prism

Helpful

Hints

Surface Area of a Rectangle Prism Formula:

SA =2 [(width × length) + (height × length) + width × height)]

✍️ *Find the surface of each prism.*

1)

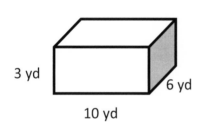

3 yd
6 yd
10 yd

2)

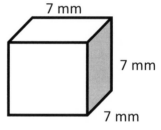

7 mm
7 mm
7 mm

3)

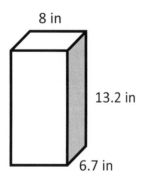

8 in
13.2 in
6.7 in

4)

17 cm

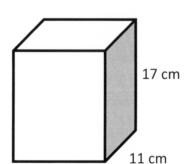

17 cm
11 cm

Volume of a Cylinder

Helpful	
Hints	Volume of Cylinder Formula = π(radius)² × height
	π = 3.14

✎ **Find the volume of each cylinder.** $(\pi = 3.14)$

1)

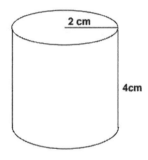

2)

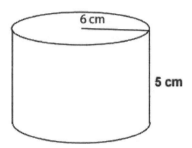

3)

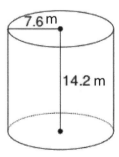

4)

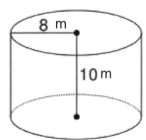

Surface Area of a Cylinder

Helpful

Hints

Surface area of a cylinder

$SA = 2\pi r^2 + 2\pi rh$

Example:

Surface area

= 1727

14 m

11 m

✎ **Find the surface of each cylinder.** ($\pi = 3.14$)

1)

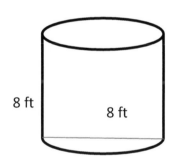

8 ft

8 ft

2)

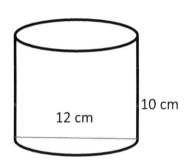

12 cm

10 cm

3)

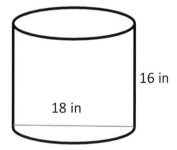

16 in

18 in

4)

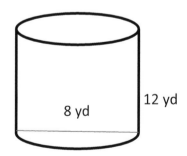

8 yd

12 yd

Answers of Worksheets – Chapter 14

Volumes of Cubes

1) 8
2) 4
3) 5
4) 36
5) 60
6) 44

Volume of Rectangle Prisms

1) 1344 cm^3
2) 1650 cm^3
3) 512 m^3
4) 1144 cm^3

Surface Area of a Cube

1) 216 mm^2
2) 486 mm^2
3) 600 cm^2
4) 384 m^2
5) 337.5 in^2
6) 766.14 ft^2

Surface Area of a Prism

1) 216 yd^2
2) 294 mm^2
3) 495.28 in^2
4) 1326 cm^2

Volume of a Cylinder

1) 50.24 cm^3
2) 565.2 cm^3
3) $2{,}575.403 \text{ m}^3$
4) 2009.6 m^3

Surface Area of a Cylinder

1) 301.44 ft^2
2) 602.88 cm^2
3) 1413 in^2
4) 401.92 yd^2

HSPT Test Review

The High School Placement Test (HSPT), also known as STS-HSPT, formulated by the Scholastic Testing Service (STS) to determine acceptance in parochial high schools.

The HSPT test consists of five multiple-choice sections:

- ✓ **Verbal Skills:** 60 questions - 16 minutes
- ✓ **Quantitative Skills:** 52 questions - 30 minutes
- ✓ **Reading:** 62 questions - 25 minutes
- ✓ **Mathematics:** 64 questions - 45 minutes
- ✓ **Language:** 60 questions - 25 minutes

Keep in mind that the Quantitative Skills and the Mathematics sections are different parts. The Quantitative Skills section contains 52 questions providing number series, geometric comparisons, and number manipulation, whereas the Mathematics part measures students' math knowledge. The Math section of the test covers arithmetic, data analysis, geometry, and algebra.

Some schools allow students to use basic calculators when taking the HSPT test.

In this section, there are two complete HSPT Mathematics Tests. Take these tests to see what score you'll be able to receive on a real HSPT test.

Good luck!

HSPT Mathematics
Practice Tests

Time to Test

Time to refine your skill with a practice examination

Take practice HSPT Math Tests to simulate the test day experience. After you've finished, score your tests using the answer keys.

Before You Start

- You'll need a pencil, a timer, and scratch papers to take the test.

- After you've finished the test, review the answer key to see where you went wrong.

- Use the answer sheet provided to record your answers. (You can cut it out or photocopy it)

- You will receive 1 point for every correct answer. There is no penalty for wrong answers.

Good Luck!

HSPT Mathematics Practice Tests Answer Sheets

Remove (or photocopy) these two answer sheets and use them to complete the practice tests.

HSPT Mathematics Practice Test 1 Answer Sheet

1	Ⓐ Ⓑ Ⓒ Ⓓ	26 Ⓐ Ⓑ Ⓒ Ⓓ	51 Ⓐ Ⓑ Ⓒ Ⓓ
2	Ⓐ Ⓑ Ⓒ Ⓓ	27 Ⓐ Ⓑ Ⓒ Ⓓ	52 Ⓐ Ⓑ Ⓒ Ⓓ
3	Ⓐ Ⓑ Ⓒ Ⓓ	28 Ⓐ Ⓑ Ⓒ Ⓓ	53 Ⓐ Ⓑ Ⓒ Ⓓ
4	Ⓐ Ⓑ Ⓒ Ⓓ	29 Ⓐ Ⓑ Ⓒ Ⓓ	54 Ⓐ Ⓑ Ⓒ Ⓓ
5	Ⓐ Ⓑ Ⓒ Ⓓ	30 Ⓐ Ⓑ Ⓒ Ⓓ	55 Ⓐ Ⓑ Ⓒ Ⓓ
6	Ⓐ Ⓑ Ⓒ Ⓓ	31 Ⓐ Ⓑ Ⓒ Ⓓ	56 Ⓐ Ⓑ Ⓒ Ⓓ
7	Ⓐ Ⓑ Ⓒ Ⓓ	32 Ⓐ Ⓑ Ⓒ Ⓓ	57 Ⓐ Ⓑ Ⓒ Ⓓ
8	Ⓐ Ⓑ Ⓒ Ⓓ	33 Ⓐ Ⓑ Ⓒ Ⓓ	58 Ⓐ Ⓑ Ⓒ Ⓓ
9	Ⓐ Ⓑ Ⓒ Ⓓ	34 Ⓐ Ⓑ Ⓒ Ⓓ	59 Ⓐ Ⓑ Ⓒ Ⓓ
10	Ⓐ Ⓑ Ⓒ Ⓓ	35 Ⓐ Ⓑ Ⓒ Ⓓ	60 Ⓐ Ⓑ Ⓒ Ⓓ
11	Ⓐ Ⓑ Ⓒ Ⓓ	36 Ⓐ Ⓑ Ⓒ Ⓓ	61 Ⓐ Ⓑ Ⓒ Ⓓ
12	Ⓐ Ⓑ Ⓒ Ⓓ	37 Ⓐ Ⓑ Ⓒ Ⓓ	62 Ⓐ Ⓑ Ⓒ Ⓓ
13	Ⓐ Ⓑ Ⓒ Ⓓ	38 Ⓐ Ⓑ Ⓒ Ⓓ	63 Ⓐ Ⓑ Ⓒ Ⓓ
14	Ⓐ Ⓑ Ⓒ Ⓓ	39 Ⓐ Ⓑ Ⓒ Ⓓ	64 Ⓐ Ⓑ Ⓒ Ⓓ
15	Ⓐ Ⓑ Ⓒ Ⓓ	40 Ⓐ Ⓑ Ⓒ Ⓓ	
16	Ⓐ Ⓑ Ⓒ Ⓓ	41 Ⓐ Ⓑ Ⓒ Ⓓ	
17	Ⓐ Ⓑ Ⓒ Ⓓ	42 Ⓐ Ⓑ Ⓒ Ⓓ	
18	Ⓐ Ⓑ Ⓒ Ⓓ	43 Ⓐ Ⓑ Ⓒ Ⓓ	
19	Ⓐ Ⓑ Ⓒ Ⓓ	44 Ⓐ Ⓑ Ⓒ Ⓓ	
20	Ⓐ Ⓑ Ⓒ Ⓓ	45 Ⓐ Ⓑ Ⓒ Ⓓ	
21	Ⓐ Ⓑ Ⓒ Ⓓ	46 Ⓐ Ⓑ Ⓒ Ⓓ	
22	Ⓐ Ⓑ Ⓒ Ⓓ	47 Ⓐ Ⓑ Ⓒ Ⓓ	
23	Ⓐ Ⓑ Ⓒ Ⓓ	48 Ⓐ Ⓑ Ⓒ Ⓓ	
24	Ⓐ Ⓑ Ⓒ Ⓓ	49 Ⓐ Ⓑ Ⓒ Ⓓ	
25	Ⓐ Ⓑ Ⓒ Ⓓ	50 Ⓐ Ⓑ Ⓒ Ⓓ	

HSPT Mathematics Practice Test 2 Answer Sheet

1	Ⓐ Ⓑ Ⓒ Ⓓ	26	Ⓐ Ⓑ Ⓒ Ⓓ	51	Ⓐ Ⓑ Ⓒ Ⓓ
2	Ⓐ Ⓑ Ⓒ Ⓓ	27	Ⓐ Ⓑ Ⓒ Ⓓ	52	Ⓐ Ⓑ Ⓒ Ⓓ
3	Ⓐ Ⓑ Ⓒ Ⓓ	28	Ⓐ Ⓑ Ⓒ Ⓓ	53	Ⓐ Ⓑ Ⓒ Ⓓ
4	Ⓐ Ⓑ Ⓒ Ⓓ	29	Ⓐ Ⓑ Ⓒ Ⓓ	54	Ⓐ Ⓑ Ⓒ Ⓓ
5	Ⓐ Ⓑ Ⓒ Ⓓ	30	Ⓐ Ⓑ Ⓒ Ⓓ	55	Ⓐ Ⓑ Ⓒ Ⓓ
6	Ⓐ Ⓑ Ⓒ Ⓓ	31	Ⓐ Ⓑ Ⓒ Ⓓ	56	Ⓐ Ⓑ Ⓒ Ⓓ
7	Ⓐ Ⓑ Ⓒ Ⓓ	32	Ⓐ Ⓑ Ⓒ Ⓓ	57	Ⓐ Ⓑ Ⓒ Ⓓ
8	Ⓐ Ⓑ Ⓒ Ⓓ	33	Ⓐ Ⓑ Ⓒ Ⓓ	58	Ⓐ Ⓑ Ⓒ Ⓓ
9	Ⓐ Ⓑ Ⓒ Ⓓ	34	Ⓐ Ⓑ Ⓒ Ⓓ	59	Ⓐ Ⓑ Ⓒ Ⓓ
10	Ⓐ Ⓑ Ⓒ Ⓓ	35	Ⓐ Ⓑ Ⓒ Ⓓ	60	Ⓐ Ⓑ Ⓒ Ⓓ
11	Ⓐ Ⓑ Ⓒ Ⓓ	36	Ⓐ Ⓑ Ⓒ Ⓓ	61	Ⓐ Ⓑ Ⓒ Ⓓ
12	Ⓐ Ⓑ Ⓒ Ⓓ	37	Ⓐ Ⓑ Ⓒ Ⓓ	62	Ⓐ Ⓑ Ⓒ Ⓓ
13	Ⓐ Ⓑ Ⓒ Ⓓ	38	Ⓐ Ⓑ Ⓒ Ⓓ	63	Ⓐ Ⓑ Ⓒ Ⓓ
14	Ⓐ Ⓑ Ⓒ Ⓓ	39	Ⓐ Ⓑ Ⓒ Ⓓ	64	Ⓐ Ⓑ Ⓒ Ⓓ
15	Ⓐ Ⓑ Ⓒ Ⓓ	40	Ⓐ Ⓑ Ⓒ Ⓓ		
16	Ⓐ Ⓑ Ⓒ Ⓓ	41	Ⓐ Ⓑ Ⓒ Ⓓ		
17	Ⓐ Ⓑ Ⓒ Ⓓ	42	Ⓐ Ⓑ Ⓒ Ⓓ		
18	Ⓐ Ⓑ Ⓒ Ⓓ	43	Ⓐ Ⓑ Ⓒ Ⓓ		
19	Ⓐ Ⓑ Ⓒ Ⓓ	44	Ⓐ Ⓑ Ⓒ Ⓓ		
20	Ⓐ Ⓑ Ⓒ Ⓓ	45	Ⓐ Ⓑ Ⓒ Ⓓ		
21	Ⓐ Ⓑ Ⓒ Ⓓ	46	Ⓐ Ⓑ Ⓒ Ⓓ		
22	Ⓐ Ⓑ Ⓒ Ⓓ	47	Ⓐ Ⓑ Ⓒ Ⓓ		
23	Ⓐ Ⓑ Ⓒ Ⓓ	48	Ⓐ Ⓑ Ⓒ Ⓓ		
24	Ⓐ Ⓑ Ⓒ Ⓓ	49	Ⓐ Ⓑ Ⓒ Ⓓ		
25	Ⓐ Ⓑ Ⓒ Ⓓ	50	Ⓐ Ⓑ Ⓒ Ⓓ		

HSPT Mathematics

Practice Test 1

- ○ **64 questions**

- ○ **Total time for this section:** 45 Minutes

- ○ **Calculators are not allowed at the test.**

1) $18a + 22 = 40, a = ?$

    A. 1

    B. 4

    C. 11

    D. 12

2) If $x = 4$, then $\dfrac{2^3}{x} =$

    A. 4

    B. 1

    C. 2

    D. 8

3) A trash container, when empty, weighs 35 pounds. If this container is filled with a load of trash that weighs 240 pounds, what is the total weight of the container and its contents?

    A. 224 pounds

    B. 275 pounds

    C. 285 pounds

    D. 325 pounds

4) Solve for a: $10x - 5 = 25$

    A. 10

    B. 5

    C. 3

    D. 2

5) Which of the following is not a multiple of 3?

    A. 13

    B. 24

    C. 27

    D. 63

6) If a rectangle is 30 feet by 45 feet, what is its area?

    A. 1,350

    B. 1,250

    C. 1,000

    D. 750

7) What is the prime factorization of 560?

   A. $2 \times 2 \times 5 \times 7$ C. 2×7

   B. $2 \times 2 \times 2 \times 2 \times 5 \times 7$ D. $2 \times 2 \times 2 \times 5 \times 7$

8) A writer finishes 180 pages of his manuscript in 20 hours. How many pages is his average?

   A. 18   C. 12

   B. 15   D. 9

9) If x is 25% percent of 250, what is x?

   A. 35   C. 62.5

   B. 55.5   D. 150

10) Find the average of the following numbers: 17, 13, 7, 21, 22

   A. 18   C. 16

   B. 17   D. 11

11) The first four terms in a sequence are shown below. What is the fifth term in the sequence?

$$\{2, 4, 8, 14, ..\}$$

   A. 14   C. 22

   B. 16   D. 34

12) $(a^2) \cdot (a^3) = $ ____

   A. a^{23} C. $2a^3$

   B. a^6 D. a^5

13) Karen is 9 years older than her sister Michelle, and Michelle is 4 years younger than her brother David. If the sum of their ages is 82, how old is Michelle?

    A. 29   C. 24

    B. 25   D. 23

14) $0.87 + 1.4 + 3.23 = ?$

    A. 3.2   C. 5.5

    B. 4.2   D. 6.63

15) Which of the following is a prime number?

    A. 8   C. 21

    B. 14   D. 37

16) The sum of two numbers is N. If one of the numbers is 6, then what is three times the other number?

    A. $3N$ C. $3(N + 6)$

    B. $3(N - 6)$ D. $(N - 2)$

17) A circle has a diameter of 3.8 inches. What is its approximate circumference?

    A. 10   C. 15

    B. 12   D. 20

18) If $x = \dfrac{2}{3}$ then $\dfrac{1}{x} = ?$

    A. $\dfrac{2}{3}$ C. $\dfrac{3}{2}$

    B. $\dfrac{1}{3}$ D. $\dfrac{1}{2}$

19) Convert 0.12 to a percent.

    A. 0.012%

    B. 0.12%

    C. 12%

    D. 1.2%

20) In the simplest form, $\frac{26}{14}$ is

    A. $\frac{3}{7}$

    B. $\frac{7}{3}$

    C. $\frac{13}{7}$

    D. $\frac{7}{13}$

21) What is the value of 6! ?

    A. 820

    B. 720

    C. 120

    D. 90

22) What number belongs in the box? $8 + \boxed{} = 3$

    A. 5

    B. 3

    C. -5

    D. 11

23) $\sqrt{49}$ is equal to:

    A. 7

    B. 8

    C. 9

    D. 14

24) Which of the following is the equivalent of 3^4?

    A. $3 \times 3 \times 3$

    B. $4 \times 4 \times 4$

    C. 27

    D. 81

25) Which of the following fractions is the largest?

A. $\frac{3}{4}$

C. $\frac{7}{9}$

B. $\frac{2}{5}$

D. $\frac{2}{3}$

26) $(5 + 7) \div (3^2 \div 3) =$ ___

A. 12

C. 4

B. $\frac{5}{7}$

D. 6

27) What's the next number in the series {20, 17, 14, 11, ?}

A. 8

C. 10

B. 9

D. 14

28) Evaluate $2x + 12$, when $x = -2$

A. 16

C. 8

B. 12

D. 6

29) If $8 < x \le 10$, then x cannot be equal to:

A. 8.5

C. 9.5

B. 9

D. 10.5

30) If $x + y = 12$, what is the value of $8x + 8y$?

A. 192

C. 104

B. 148

D. 96

31) If $6 + x \geq 18$, then $x \geq$?

    A. 3

    B. 6

    C. 12

    D. $18x$

32) I've got 34 quarts of milk and my family drinks 2 gallons of milk per week. How many weeks will that last us?

    A. 2 weeks

    B. 2.5 weeks

    C. 3.25 weeks

    D. 4.25 weeks

33) If $-8a = 64$, then $a =$ ___

    A. -8

    B. 8

    C. 16

    D. 0

34) Factor this equation: $x^2 + 5x - 6$

    A. $x^2(5 + 6)$

    B. $x(x + 5 - 6)$

    C. $(x + 6)(x - 1)$

    D. $(x + 6)(x - 6)$

35) $126 \times 45 =$?

    A. 5,760

    B. 5,670

    C. 5,607

    D. 5,067

36) If A = 6, B = 4 and C = 5, then 3ABC = ?

    A. 120

    B. 18

    C. 645

    D. 360

37) A baker uses 4 eggs to bake a cake. How many cakes will he be able to bake with 188 eggs?

    A. 46   C. 48

    B. 47   D. 49

38) Which of the following is the product of 19 and 12?

    A. 31   C. 228

    B. 7   D. 1.59

39) What is the difference between 152 and 83?

    A. 235   C. 12,616

    B. 69   D. 1.8

40) Which of the following is equal to 2.5?

    A. 3   C. $\frac{25}{10}$

    B. $2\frac{1}{10}$ D. $\frac{25}{100}$

41) What is the sum of $\frac{7}{12} + \frac{4}{3} + \frac{2}{6}$?

    A. 2.1   C. $2\frac{1}{4}$

    B. 2   D. 1

42) The number 0.9 can also represented by which of the following?

A. $\dfrac{9}{10}$

C. $\dfrac{9}{1,000}$

B. $\dfrac{9}{100}$

D. $\dfrac{9}{10,000}$

43) Which of the following is the correct calculation for 4!?

A. 4×3

C. $1 \times 2 \times 3 \times 4 \times 5$

B. $1 \times 2 \times 3$

D. $4 \times 3 \times 2 \times 1$

44) $\dfrac{13}{25}$ is equal to:

A. 5.2

C. 0.05

B. 0.52

D. 0.5

45) Julie gives 8 pieces of candy to each of her friends. If Julie gives all her candy away, which amount of candy could have been the amount she distributed?

A. 187

C. 243

B. 216

D. 223

46) How much greater is the value of $5x + 8$ than the value of $5x - 3$?

A. 7

C. 11

B. 9

D. 13

47) A woman weighs 135 pounds. She gains 15 pounds one month and 8 pounds the next month. What is her new weight?

A. 152 Pounds

C. 158 Pounds

B. 146 Pounds

D. 138 Pounds

48) $(x + 2)(x + 3) = ?$

    A. $x^2 - 3x + 6$

    B. $x^2 + 5x + 6$

    C. $x^2 + 6x + 5$

    D. $x^2 + 6x - 5$

49) What is the perimeter of the triangle in the provided diagram?

    A. 25

    B. 15,625

    C. 75

    D. 625

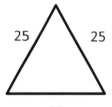

50) The equation of a line is given as : $y = 5x - 3$. Which of the following points does not lie on the line?

    A. (1, 2)

    B. (-2, -13)

    C. (3, 18)

    D. (2, 7)

51) How long is the line segment shown on the number line below?

    A.

    B.

    C.

    D. 9

52) $x^2 - 81 = 0$, x could equal:

    A. -6

    B. -9

    C. 12

    D. 81

53) $(x^8)^2 = ?$

    A. $2x^8$ C. x^{16}

    B. x^{28} D. x^{10}

54) In the diagram provided, what is the value of b?

    A. 180°

    B. 120°

    C. 135°

    D. 45°

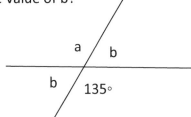

55) If a rectangular swimming pool has a perimeter of 112 feet and is 22 feet wide, what is its area?

    A. 1,496   C. 2,464

    B. 90   D. 748

56) What is $\sqrt{25} \times \sqrt{16}$

    A. $4\sqrt{5}$ C. 20

    B. 41   D. $\sqrt{41}$

57) The cube of 8 is ___ .

    A. 512   C. 80

    B. 64   D. 24

58) What's the next number in the series {29, 24, 19, 14, ?}

    A. 14

    B. 9

    C. 4

    D. 19

59) What's the greatest common factor of the 25 and 15?

    A. 44

    B. 15

    C. 5

    D. 3

60) What is 8,923.2769 rounded to the nearest tenth?

    A. 8923.3

    B. 8923.277

    C. 8923

    D. 8923.27

61) What is the radicand in $\sqrt[3]{216}$?

    A. 3

    B. 72

    C. 36

    D. 6

62) If $x = 7$ what's the value of $6x^2 + 5x - 13$?

    A. 64

    B. 316

    C. 416

    D. 293

63) Line l passes through the point (−1, 2). Which of the following CANNOT be the equation of line l?

    A. $y = 1 - x$

    B. $y = x + 1$

    C. $x = -1$

    D. $y = x + 3$

64) The circumference of a circle is 30 cm. what is the approximate radius of the circle?

    A. 2.4 cm   C. 8.0 cm

    B. 4.8 cm   D. 9.5 cm

IF YOU FINISH BEFORE TIME IS CALLED, YOU MAY CHECK YOUR WORK ON THIS TEST.   STOP

HSPT Mathematics

Practice Test 2

- ○ **64 questions**
- ○ **Total time for this section:** 45 Minutes
- ○ **Calculators are not allowed at the test.**

1) Which of the following is the equivalent of 4^3?

A. 4×4

C. 64

B. $4 \times 4 \div 3$

D. 40

2) In the following figure, the two lines are perpendicular and the vertical line is the diameter of the circle. What is the value of x?

A. 5

B. 10

C. 12

D. 14

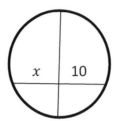

3) The sum of 8 numbers is greater than 240 and less than 320. Which of the following could be the average (arithmetic mean) of the numbers?

A. 30

C. 40

B. 35

D. 45

4) Which of the following fractions is the largest?

A. $\dfrac{5}{8}$

C. $\dfrac{8}{9}$

B. $\dfrac{3}{7}$

D. $\dfrac{5}{11}$

5) $-18 + 6 \times (-5) - [4 + 22 \times (-4)] \div 2 = ?$

A. - 6

C. -1

B. $\dfrac{3}{4}$

D. 6

6) What's the next number in the series {32, 29, 26, 23, ?}

   A. 15   C. 17

   B. 20   D. 13

7) Evaluate 5a + 25, when a = -5

   A. -150   C. 50

   B. 30   D. 0

8) $12a + 20 = 140$, $a = ?$

   A. 12   C. 14

   B. 10   D. 18

9) During a fund-raiser, each of the 45 members of a group sold candy bars. If each member sold an average of five candy bars, how many total bars did the group sell?

   A. 21   C. 195

   B. 56   D. 225

10) Which of the following is a multiple of 4?

   A. 38   C. 85

   B. 46   D. 104

11) Which of the following is the product of 12 and 4?

   A. 94   C. 24

   B. 48   D. 16

12) What's the greatest common factor of the 18 and 32?

    A. 11   C. 2

    B. 12   D. 4

13) What's the least common multiple (LCM) of 8 and 14?

    A. 8 AND 14 HAVE NO COMMON   C. 112
    MULTIPLES
                                            D. 56
    B. 96

14) How many $\frac{1}{4}$ pound paperback books together weigh 30 pounds?

    A. 80   C. 105

    B. 95   D. 120

15) A woman owns a dog walking business. If 3 workers can walk 9 dogs, how many dogs can 5 workers walk?

    A. 13   C. 15

    B. 17   D. 19

16) What's the greatest common factor of the 18, 20 and 38?

    A. 5   C. 3

    B. 4   D. 2

17) Emily and Daniel have taken the same number of photos on their school trip. Emily has taken 5 times as many as photos as Claire and Daniel has taken 16 more photos than Claire. How many photos has Claire taken?

    A. 4   C. 8

    B. 6   D. 10

18) The distance between cities A and B is approximately 2,600 miles. If you drive an average of 68 miles per hour, how many hours will it take you to drive from city A to city B?

A. approximately 41 hours

C. approximately 29 hours

B. approximately 38 hours

D. approximately 27 hours

19) If 6 garbage trucks can collect the trash of 36 homes in a day. How many trucks are needed to collect in 180 houses?

A. 18

C. 15

B. 19

D. 30

20) If $x = 6$, then $\dfrac{6^5}{x} =$

A. 30

C. 1,296

B. 7,776

D. 96

21) Solve for a: 8a − 5 = 11

A. 64

C. 31.42

B. 0.45

D. 2

22) Which of the following is not a multiple of 5?

A. 12

C. 15

B. 30

D. 20

23) $\dfrac{(15\ feet\ +7\ yards)}{4} =$ ____

   A. 9 feet

   B. 7 feet

   C. 28 feet

   D. 4 feet

24) Sylvia is 7 years older than her sister Danna, and Danna is 5 years younger than their brother Jerry. If the sum of their ages is 72, how old is Danna?

   A. 18

   B. 22

   C. 26

   D. 20

25) $(p^4) \cdot (p^5) =$ ____

   A. p^{20}

   B. $2p^9$

   C. p^9

   D. $2p^{20}$

26) If $(4.2 + 4.3 + 4.5)x = x$, then what is the value of x?

   A. 0

   B. $\dfrac{1}{10}$

   C. 1

   D. 10

27) While at work, Emma checks her email once every 90 minutes. In 9-hour, how many times does she check her email?

   A. 4 Times

   B. 5 Times

   C. 7 Times

   D. 6 Times

28) Which of the following is a prime number?

   A. 10

   B. 8

   C. 11

   D. 93

29) The sum of two numbers is X. if one of the numbers is 9, then two times the other number would be what?

A. 2X

B. 2 + X × 2

C. 2(X + 9)

D. 2(X - 9)

30) A circle has a diameter of 8 inches. What is its approximate circumference?

A. 6.28

B. 25.12

C. 34.85

D. 35.12

31) If $x = \frac{5}{7}$ then $\frac{1}{x} = ?$

A. $\frac{7}{5}$

B. $\frac{5}{7}$

C. 5

D. 7

32) What is 5231.48245 rounded to the nearest tenth?

A. 5231.482

B. 5231.5

C. 5231

D. 5231.48

33) Which of the following is NOT a factor of 50?

A. 5

B. 2

C. 10

D. 15

34) What is the place value of 2 in 5.7325?

A. hundredths

B. thousandths

C. ten thousandths

D. hundred thousandths

35) Which symbol belongs in the circle? 0.632 ◯ 0.0540

    A. <

    B. >

    C. =

    D. ≤

36) Convert 0.025 to a percent.

    A. 0.03%

    B. 0.25%

    C. 2.50%

    D. 25%

37) In the simplest form, $\frac{18}{24}$ is

    A. $\frac{2}{3}$

    B. $\frac{3}{2}$

    C. $\frac{4}{3}$

    D. $\frac{3}{4}$

38) $\frac{7}{25}$ is equal to:

    A. 0.3

    B. 2.8

    C. 0.03

    D. 0.28

39) What number belong in the box? - 2 + ☐ = 0

    A. -2

    B. 0

    C. 2

    D. -1

40) What is the square root of 64?

    A. 6

    B. 8

    C. 9

    D. 4,096

41) If $6.5 < x \le 9.0$, then x cannot be equal to:

A. 6.5

C. 8.5

B. 7.2

D. 9

42) 8 feet, 10 inches + 5 feet, 12 inches = how many inches?

A. 178 inches

C. 182 inches

B. 188 inches

D. 200 inches

43) If a = 4 then $a^a \cdot a$ = ?

A. 40

C. 1,024

B. 256

D. $4x$

44) Use the diagram provided as a reference. If the length between point A and C is 67, and the length between point A and B is 26, what is the length between point B and C?

A. 41

B. 93

C. 31

D. 46

45) If $7 + 2x \le 15$, then $x \le$?

A. $14x$

C. -4

B. 4

D. $15x$

46) What is the sum of $\frac{1}{3} + \frac{2}{5} + \frac{1}{2}$?

A. 0.9

C. 1.23

B. 1

D. 2

47) If Ella needed to buy 6 bottles of soda for a party in which 10 people attended, how many bottles of soda will she need to buy for a party in which 5 people are attending?

A. 3

C. 9

B. 6

D. 12

48) Which of the following represents the reduced form for 1.8?

A. $1\frac{8}{10}$

C. $\frac{18}{10}$

B. $1\frac{4}{5}$

D. $\frac{36}{20}$

49) What is the difference between 86 and 10?

A. 76

C. 860

B. 96

D. 8.6

50) The fraction $\frac{3}{4}$ can also be written as which of the following?

A. $\frac{4}{3}$

C. 0.75

B. 0.34

D. 34.75

51) Which of the following is equivalent to $7 \times 7 \times 7 \times 7$?

A. $\sqrt{7} \times 7$

C. $7 \div 7$

B. 7^4

D. $7,000$

52) Emily lives $5\frac{1}{4}$ miles from where she works. When traveling to work, she walks to a bus stop $\frac{1}{3}$ of the way to catch a bus. How many miles away from her house is the bus stop?

A. $4\frac{1}{3}$ Miles

C. $2\frac{3}{4}$ Miles

B. $4\frac{3}{4}$ Miles

D. $1\frac{3}{4}$ Miles

53) A bread recipe calls for $3\frac{1}{3}$ cups of flour. If you only have $2\frac{5}{6}$ cups, how much more flour is needed?

A. 1

C. 2

B. $\frac{1}{2}$

D. $\frac{5}{6}$

54) If $l = 2$, $a = 4$ and $b = 3$, then $4lab = ?$

A. 24

C. 96

B. 48

D. 144

55) Mario loaned Jett $1,200 at a yearly interest rate of 5%. After one year what is the interest owned on this loan?

A. $1,260

C. $60

B. $600

D. $26

56) 235 × 34 = ?

  A. 7,099

  B. 7,909

  C. 7,990

  D. 7,009

57) Charlotte is 46 years old, twice as old as Avery. How old is Avery?

  A. 23 years old

  B. 28 years old

  C. 20 years old

  D. 18 years old

58) In the given diagram, the height is 9 cm. what is the area of the triangle?

  A. 23 cm^2

  B. 46 cm^2

  C. 126 cm^2

  D. 252 cm^2

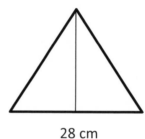

28 cm

59) In the figure below, line A is parallel to line B. what is the value of angle x?

  A. 35 degree

  B. 45 degree

  C. 100 degree

  D. 145 degree

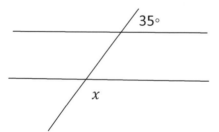

60) What is the value of 5!

  A. 120

  B. 15

  C. 12

  D. 10

61) There are two pizza ovens in a restaurant. Oven 1 burns four times as many pizzas as oven 2. If the restaurant had a total of 15 burnt pizzas on Saturday, how many pizzas did oven 2 burn?

A. 3

B. 6

C. 9

D. 12

62) Which of the following is the correct calculation for 7!?

A. $7 \times 6 \times 5 \times 4 \times 3 \times 2 \times 1 \times 0$

B. $1 \times 2 \times 3 \times 4 \times 5 \times 6$

C. $1 \times 2 \times 3 \times 4 \times 5 \times 6 \times 7$

D. $1 \times 2 \times 3 \times 4 \times 5 \times 6 \times 7 \times 8$

63) Jacob is having a birthday party for his girl and is serving orange juice to the 8 children in attendance. If Jacob has 1 liter of orange juice and wants to divide it equally among the children, how many liters does each child get?

A. $\frac{1}{8}$

B. $\frac{1}{7}$

C. $\frac{1}{9}$

D. $\frac{1}{16}$

64) The number 0.04 can also represented by which of the following?

A. $\frac{4}{10}$

B. $\frac{4}{100}$

C. $\frac{4}{1,000}$

D. $\frac{4}{10,000}$

IF YOU FINISH BEFORE TIME IS CALLED, YOU MAY CHECK YOUR WORK ON THIS TEST.   STOP

HSPT Mathematics Practice Tests Answers and Explanations

HSPT Mathematics Practice Test 1 Answers

1-	A	17-	B	33-	A	49-	C
2-	C	18-	C	34-	C	50-	C
3-	B	19-	C	35-	B	51-	D
4-	C	20-	C	36-	D	52-	B
5-	A	21-	B	37-	B	53-	C
6-	A	22-	C	38-	C	54-	D
7-	B	23-	A	39-	B	55-	D
8-	D	24-	D	40-	C	56-	C
9-	C	25-	C	41-	C	57-	A
10-	C	26-	C	42-	A	58-	B
11-	C	27-	A	43-	D	59-	C
12-	D	28-	C	44-	B	60-	A
13-	D	29-	D	45-	B	61-	D
14-	C	30-	D	46-	C	62-	B
15-	D	31-	C	47-	C	63-	B
16-	B	32-	D	48-	B	64-	B

HSPT Mathematics Practice Test 2 Answers

1-	C	17-	A	33-	D	49-	A
2-	C	18-	B	34-	B	50-	C
3-	B	19-	D	35-	B	51-	B
4-	C	20-	C	36-	C	52-	D
5-	A	21-	D	37-	D	53-	D
6-	B	22-	A	38-	D	54-	C
7-	D	23-	A	39-	C	55-	C
8-	B	24-	D	40-	B	56-	C
9-	D	25-	C	41-	A	57-	A
10-	D	26-	A	42-	A	58-	C
11-	B	27-	D	43-	C	59-	D
12-	C	28-	C	44-	A	60-	A
13-	D	29-	D	45-	B	61-	A
14-	D	30-	B	46-	C	62-	C
15-	C	31-	A	47-	A	63-	A
16-	D	32-	B	48-	B	64-	B

HSPT Mathematics Practice Tests 1 Answers and Explanations

1) Choice A is correct

$$18a + 22 = 40 \rightarrow 18a = 40 - 22 \rightarrow 18a = 18 \rightarrow a = 1$$

2) Choice C is correct

$$\frac{2^3}{x} = \frac{2^3}{4} = \frac{8}{4} = 2$$

3) Choice B is correct

$$240 + 35 = 275$$

4) Choice C is correct

$$10x - 5 = 25$$

$$10x = 30$$

$$x = \frac{30}{10} = 3$$

5) Choice A is correct

A. $\frac{13}{3} = 4.33$

B. $\frac{24}{3} = 8$

C. $\frac{12}{3} = 4$

D. $\frac{27}{3} = 9$

6) Choice A is correct

Area of a rectangle = width × length = 30 × 45 = 1,350

7) Choice B is correct

Find the value of each choice:

A. 2 × 2 × 5 × 7 = 140

B. 2 × 2 × 2 × 2 × 5 × 7 = 560

C. 2 × 7 = 14

D. 2 × 2 × 2 × 5 × 7 = 280

8) Choice D is correct

180 ÷ 20 = 9

9) Choice C is correct

$\frac{25}{100}$ 250 = x

$x = \frac{25 \times 250}{100}$ = 62.5

10) Choice C is correct

$\frac{17 + 13 + 7 + 21 + 22}{5} = \frac{80}{5}$ = 16

11) Choice C is correct

The difference of 2 and 4 is 2. The difference of 4 and 8 is 4 and the difference of 8 and 14 is 6. Then, the difference of 14 and the next number should be 8. The answer is 22. 14 + 8 = 22

12) Choice D is correct

$(a^2) \cdot (a^3) = a^{2+3} = a^5$

13) Choice D is correct

Michelle = Karen − 9

Michelle = David − 4

Karen + Michelle + David = 82

Karen + 9 = Michelle ⟹ Karen = Michelle − 9

Karen + Michelle + David = 82

Now, replace the ages of Karen and David by Michelle. Then:

Michelle + 9 + Michelle + Michelle + 4 = 82

3Michelle + 13 = 82 ⟹ 3Michelle = 82 − 13

3Michelle = 69

Michelle = 23

14) Choice C is correct

0.87 + 1.4 + 3.23 = 5.5

15) Choice D is correct

A. 8 is not prime. It is divisible by 2.

B. 14 is not prime. It is divisible by 2.

C. 21 is not prime. It is divisible by 3.

D. 37 is prime!

16) Choice B is correct

Let x and y be the numbers. Then:

$$x + y = N, x = 6 \rightarrow 6 + y = N \rightarrow y = N - 6, 3y = 3(N - 6)$$

17) Choice B is correct

Circumference = πd ⟹ C = 3.8π = 11.932 ≅ 12

18) Choice C is correct

$$\frac{\frac{1}{2}}{\frac{2}{3}} = \frac{3}{2}$$

19) Choice C is correct

$0.12 \times 100 = 12\%$

20) Choice C is correct

$\dfrac{26}{14} = \dfrac{13}{7}$

21) Choice B is correct

$6! = 6 \times 5 \times 4 \times 3 \times 2 \times 1 = 720$

22) Choice C is correct

$8 + x = 3$

$x = 3 - 8$

$x = -5$

23) Choice A is correct

$\sqrt{49} = 7$

24) Choice D is correct

$3^4 = 3 \times 3 \times 3 \times 3 = 81$

25) Choice C is correct

A. $\dfrac{3}{4} = 0.75$

B. $\dfrac{2}{5} = 0.4$

C. $\dfrac{7}{9} = 0.77$

D. $\dfrac{2}{3} = 0.66$

26) Choice C is correct

$(5 + 7) \div (3^2 \div 3) = (12) \div (3) = 4$

27) Choice A is correct

The difference of each number and next number is 3.

{20, 17, 14, 11, 8}

28) Choice C is correct

$$2x + 12 = 2(-2) + 12 = -4 + 12 = 8$$

29) Choice D is correct

$8 < x \leq 10$, then x cannot be equal to 10.5 .

30) Choice D is correct

$x + y = 12$

$8x + 8y = 8(x + y) \rightarrow 12 \times 8 = 96$

31) Choice C is correct

$6 + x \geq 18 \rightarrow x \geq 18 - 6 \qquad \rightarrow \qquad x \geq 12$

32) Choice D is correct

1 quart = 0.25 gallon

34 quarts = 34 × 0.25 = 8.5 gallons

then: $\frac{8.5}{2}$ = 4.25 weeks

33) Choice A is correct

$-8a = 64$

$a = \frac{64}{-8} = -8$

34) Choice C is correct

$x^2 + 5x - 6 = (x + 6)(x - 1)$

35) Choice B is correct

$126 \times 45 = 5{,}670$

36) Choice D is correct

$3ABC = 3(6)(4)(5) = 360$

37) Choice B is correct

Write a proportion and solve.

$\frac{4}{1} = \frac{188}{x} \qquad \rightarrow x = \frac{188}{4} = 47$

38) Choice C is correct

$19 \times 12 = 228$

39) Choice B is correct

$152 - 83 = 69$

40) Choice C is correct

$\frac{25}{10} = 2.5$

41) Choice C is correct

$\frac{7}{12} + \frac{4}{3} + \frac{2}{6} = \frac{7+4(4)+2(2)}{12} = \frac{27}{12} = \frac{9}{4} = 2\frac{1}{4}$

42) Choice A is correct

$\frac{9}{10} = 0.9$

43) Choice D is correct

$4! = 4 \times 3 \times 2 \times 1$

44) Choice B is correct

$\frac{13}{25} = 0.52$

45) Choice B is correct

Since Julie gives 8 pieces of candy to each of her friends, then, then number of pieces of candies must be divisible by 8.

A. $187 \div 8 = 23.375$
B. $216 \div 8 = 27$
C. $343 \div 8 = 42.875$
D. $223 \div 8 = 27.875$

Only choice B gives a whole number.

46) Choice C is correct

$$(5x + 8) - (5x - 3) = 5x + 8 - 5x + 3 = 11$$

47) Choice C is correct

135 + 15 + 8 = 158

48) Choice B is correct

$(x + 2)(x + 3) = x^2 + 3x + 2x + 6 = x^2 + 5x + 6$

49) Choice C is correct

Perimeter of a triangle = side 1 + side 2 + side 3 = 25 + 25 + 25 = 75

50) Choice C is correct

Let's review the choices provided. Put the values of x and y in the equation.

A. (1, 2) $\Rightarrow x = 1 \Rightarrow y = 2$ This is true!

B. (−2, −13) $\Rightarrow x = -2 \Rightarrow y = -13$ This is true!

C. (3, 18) $\Rightarrow x = 3 \Rightarrow y = 12$ This is not true!

D. (2, 7) $\Rightarrow x = 2 \Rightarrow y = 7$ This is true!

51) Choice D is correct

The line is from -8 to $+1$. The difference of these two numbers is: $1 - (-8) = 1 + 8 = 9$

52) Choice B is correct

$x^2 - 81 = 0$

$x^2 = 81$

$x = 9 \text{ or } - 9$

53) Choice C is correct

$(x^8)^2 = x^{16}$

54) Choice D is correct

Angle b and angle 135 are supplementary angles. (their sum is 180 degrees)

Angle b = 180 − 135 = 45

55) Choice D is correct

Perimeter of rectangles = $2(width + length)$

Area of a rectangles = $width \cdot length$

Perimeter of rectangles = $2(width + length) \rightarrow 112 = 2(22 + length) \rightarrow$

$$112 = 44 + 2(length) \rightarrow 68 = 2(length) \rightarrow length = 34$$

Area of a rectangles = $width \cdot length =$ = 22 × 34 = 748

56) Choice C is correct

$\sqrt{25} \times \sqrt{16}$ = 5 × 4 = 20

57) Choice A is correct

8 × 8 × 8 = 512

58) Choice B is correct

The difference between any number and its next number is 5. {29, 24, 19, 14, 9}

59) Choice C is correct

The factors of 15 are: { 1, 3, 5, 15}

The factors of 25 are: {1, 5, 25}

Then: Greatest Common Factor = 5

60) Choice A is correct

The tenth value is 2. 8923.27 is closer to 8923.3 than 8923.2

61) Choice D is correct

$\sqrt[3]{216} = \sqrt[3]{6^3} = 6$

62) Choice B is correct

Plug in the value of x in the expression. Then: $6x^2 + 5x - 13 = 6(7)^2 + 5(7) - 13 = 316$

63) Choice B is correct

Plug in the values of x and y in the equations provided:

A. $y = 1 - x$ → $y = 1 - (-1) = 2$ This is true!

B. $y = x + 1$ → $y = (-1) + 1 = 0$ This is NOT true!

C. $x = -1$ → $(-1) = -1$ This is true!

D. $y = x + 3$ → $y = (-1) + 3 = 2$ This is true!

64) Choice B is correct

Use circle circumference formula:

C = 2πr

30 = 2πr

$r = \dfrac{30}{2\pi} = 4.77 \cong 4.8$

HSPT Mathematics Practice Tests 2 Answers and Explanations

1) Choice C is correct

$4^3 = 4 \times 4 \times 4 = 64$

2) Choice C is correct

Since the lines are perpendicular and the vertical line is the diameter of the circle, two sides of the horizontal line are equal.

3) Choice B is correct

The sum of numbers divided by 8 give the average of numbers. Then:

$$\frac{240}{8} < x < \frac{320}{8}$$

$30 < x < 40$

From choices provided, only 35 is correct.

4) Choice C is correct

A. $\dfrac{5}{8} = 0.625$

B. $\dfrac{3}{7} = 0.43$

C. $\dfrac{8}{9} = 0.88$

D. $\dfrac{5}{11} = 0.45$

5) Choice A is correct

Use PEMDAS (order of operation):

$$-18 + 6 \times (-5) - [4 + 22 \times (-4)] \div 2 = -18 - 30 - [4 - 88] \div 2 = -48 - [-84] \div 2$$
$$= -48 + 84 \div 2 = -48 + 42 = -6$$

6) Choice B is correct

The difference of each number and the next number is 3. {32, 29, 26, 23, 20}

7) Choice D is correct

$$5a + 25$$

$$5(-5) + 25 = 0$$

8) Choice B is correct

$$12a + 20 = 140 \rightarrow 12a = 140 - 20 \rightarrow 12a = 120 \rightarrow a = 10$$

9) Choice D is correct

$45 \times 5 = 225$

10) Choice D is correct

A. $\dfrac{38}{4} = 9.5$

B. $\dfrac{46}{4} = 11.5$

C. $\dfrac{85}{4} = 21.25$

D. $\dfrac{104}{4} = 26$

11) Choice B is correct

$12 \times 4 = 48$

12) Choice C is correct

The factors of 18 are: { 1, 2, 3, 6, 9, 18}

The factors of 32 are: { 1, 2, 4, 8, 16, 32}

GCF = 2

13) Choice D is correct

The smallest number that is divisible by both 8 and 14 is 56. LCM = 56

14) Choice D is correct

$$30 \div \frac{1}{4} = 120$$

15) Choice C is correct

Each worker can walk 3 dogs: 9 ÷ 3 = 3

5 workers can walk 15 dogs. 5 × 3 = 15

16) Choice D is correct

The factors of 18 are: {1, 2, 3, 6, 9, 18}

The factors of 20 are: {1, 2, 4, 5, 10, 20}

The factors of 38 are: {1, 2, 19, 38}

GCF = 2

17) Choice A is correct

Emily = Daniel, Emily = 5 (Claire)

Daniel = 16 + Claire

Emily = Daniel   →   Emily = 16 + Claire

Emily = 5 (Claire) →   5 (Claire) = 16 + Claire →   5 (Claire) – Claire = 16

4 (Claire) = 16, Claire = 4

18) Choice B is correct

Speed = $\frac{distance}{time}$

$68 = \frac{2,600}{time}$ $\rightarrow time = \frac{2,600}{68} = 38.23$

19) Choice D is correct

Write a proportion and solve.

$$\frac{6}{36} = \frac{x}{180} \rightarrow x = \frac{6 \times 180}{36} = 30$$

20) Choice C is correct

$$\frac{6^5}{6} = 6^4 = 1,296$$

21) Choice D is correct

$$8a - 5 = 11, 8a = 11 + 5, 8a = 16$$

$$a = 2$$

22) Choice A is correct

A. $\frac{12}{5} = 2.4$

B. $\frac{30}{5} = 6$

C. $\frac{15}{5} = 3$

D. $\frac{20}{5} = 4$

23) Choice A is correct

$$\frac{(15\ feet + 7\ yards)}{4} = \frac{(15\ feet + 21\ feet)}{4} = \frac{(36\ feet)}{4} = 9\ feet$$

24) Choice D is correct

Sylvia = Danna + 7

Danna = Jerry − 5 → Jerry = Danna + 5

Jerry + Danna + Sylvia = 72

(Danna + 5) + Danna + (Danna + 7) = 72

3 Danna + 12 = 72

3 Danna = 60, Danna = 20

25) Choice C is correct

$(p^4) \cdot (p^5) = p^{4+5} = p^9$

26) Choice A is correct

$(4.2 + 4.3 + 4.5)\, x = x$

$13x = x$

Then $x = 0$

27) Choice D is correct

9 hour = 540 minutes

$$\frac{90}{1} = \frac{540}{x} \quad \rightarrow \quad x = \frac{540}{90} = 6$$

28) Choice C is correct

A. 10 $\rightarrow$ 10 is not prime. It is divisible by 2.

B. 8 $\rightarrow$ 8 is not prime. It is divisible by 2.

C. 11 $\rightarrow$ 11 is prime!

D. 93 $\rightarrow$ 93 is not prime. It is divisible by 3.

29) Choice D is correct

A + B = X

A + 9 = X

A = X − 9

2A = 2(X − 9)

30) Choice B is correct

Diameter = 8, then: Radius = 4

C = 2πr , C = 8π $\rightarrow$ C = 25.12

31) Choice A is correct

$$\frac{1}{x} = \frac{\frac{1}{1}}{\frac{5}{7}} = \frac{7}{5}$$

32) Choice B is correct

The number in the tenth place is 4. The number 5,231.48 is closer to 5,231.5 than 5,231.4

33) Choice D is correct

The factors of 50 are: { 1, 2, 5, 10, 25, 50}

34) Choice B is correct

2 is in the thousandths place.

35) Choice B is correct

0.632 is bigger than 0.0540.

0.632 > 0.0540

36) Choice C is correct

0.025 × 100 = 2.5%

37) Choice D is correct

$$\frac{18}{24} = \frac{3}{4}$$

38) Choice D is correct

$$\frac{7}{25} = 0.28$$

39) Choice C is correct

-2 + 2 = 0

40) Choice B is correct

$\sqrt{64} = 8$

41) Choice A is correct

$6.5 < x \leq 9.0$, then cannot be equal to 6.5

42) Choice A is correct

1 feet = 12 inches

8 feet, 10 inches = 106 inches

5 feet, 12 inches = 72 inches

106 + 72 = 178

43) Choice C is correct

$a^a \cdot a = (4)^4 \cdot 4 = 256 \cdot 4 = 1{,}024$

44) Choice A is correct

$67 - 26 = 41$

45) Choice B is correct

$7 + 2x \leq 15$

$2x \leq 15 - 7$

$2x \leq 8$

$x \leq 4$

46) Choice C is correct

$\frac{1}{3} + \frac{2}{5} + \frac{1}{2} = \frac{10+12+15}{30} = \frac{37}{30} = 1.23$

47) Choice A is correct

Write a proportion and solve. $\frac{6}{10} = \frac{x}{5}$ $\rightarrow$ $x = \frac{6 \times 5}{10} = 3$

48) Choice B is correct

The reduced fraction equal to 1.8 is $1\frac{4}{5}$

49) Choice A is correct

$86 - 10 = 76$

50) Choice C is correct

$\frac{3}{4} = 0.75$

51) Choice B is correct

$7 \times 7 \times 7 \times 7 = 7^4$

52) Choice D is correct

$\frac{1}{3}$ of the distance is $5\frac{1}{4}$ miles. Then: $\frac{1}{3} \times 5\frac{1}{4} = \frac{1}{3} \times \frac{21}{4} = \frac{21}{12}$

Converting $\frac{21}{12}$ to a mixed number gives: $\frac{21}{12} = 1\frac{9}{12} = 1\frac{3}{4}$

53) Choice D is correct

$2\frac{2}{3} - 1\frac{5}{6} = 2\frac{4}{6} - 1\frac{5}{6} = \frac{16}{6} - \frac{11}{6} = \frac{5}{6}$

54) Choice C is correct

$$4lab = 4(2)(4)(3) = 96$$

55) Choice C is correct

Use interest rate formula:

$$Interest = principal \times rate \times time = 1,200 \times 0.05 \times 1 = 60$$

56) Choice C is correct

$235 \times 34 = 7,990$

57) Choice A is correct

Charlotte = 46

Charlotte = 2 Avery

Avery = $\frac{46}{2}$ = 23

58) Choice C is correct

$A = \frac{1}{2} bh$

$A = \frac{1}{2} (28)(9) = 126$

59) Choice D is correct

Angle x and angle 35 degrees are supplementary angles. Then: $180° - 35° = 145°$

$x = 145°$

60) Choice A is correct

$5! = 5 \times 4 \times 3 \times 2 \times 1 = 120$

61) Choice A is correct

Oven 1 = 4 oven 2

If Oven 2 burns 3 then oven 1 burns 12 pizzas. 3 + 12 = 15

62) Choice C is correct

$7! = 7 \times 6 \times 5 \times 4 \times 3 \times 2 \times 1$

63) Choice A is correct

8 children

1 liter of orange juice

Then each child gets $\frac{1}{8}$ of litter.

64) Choice B is correct

$\frac{4}{100} = 0.04$

"Effortless Math Education" Publications

Effortless Math Education authors' team strives to prepare and publish the best quality Mathematics learning resources to make learning Math easier for all. We hope that our publications help you or your student learn Math in an effective way.

We all in Effortless Math wish you good luck and successful studies!

Effortless Math Authors

www.EffortlessMath.com

... So Much More Online!

✓ FREE Math lessons

✓ More Math learning books!

✓ Mathematics Worksheets

✓ Online Math Tutors

Need a PDF version of this book?

Please visit www.EffortlessMath.com

51868827R00125